T0297089

Studies in Computational Intelligence

Volume 614

Series editor

Janusz Kacprzyk, Polish Academy of Sciences, Warsaw, Poland
e-mail: kacprzyk@ibspan.waw.pl

About this Series

The series "Studies in Computational Intelligence" (SCI) publishes new developments and advances in the various areas of computational intelligence—quickly and with a high quality. The intent is to cover the theory, applications, and design methods of computational intelligence, as embedded in the fields of engineering, computer science, physics and life sciences, as well as the methodologies behind them. The series contains monographs, lecture notes and edited volumes in computational intelligence spanning the areas of neural networks, connectionist systems, genetic algorithms, evolutionary computation, artificial intelligence, cellular automata, self-organizing systems, soft computing, fuzzy systems, and hybrid intelligent systems. Of particular value to both the contributors and the readership are the short publication timeframe and the worldwide distribution, which enable both wide and rapid dissemination of research output.

More information about this series at http://www.springer.com/series/7092

Roger Lee

Editor

Computer and Information Science 2015

 Springer

Editor
Roger Lee
Software Engineering and Information
  Institute
Central Michigan University
Mount Pleasant, MI
USA

ISSN 1860-949X   ISSN 1860-9503   (electronic)
Studies in Computational Intelligence
ISBN 978-3-319-23466-3   ISBN 978-3-319-23467-0   (eBook)
DOI 10.1007/978-3-319-23467-0

Library of Congress Control Number: 2015950927

Printed on acid-free paper

Springer International Publishing AG Switzerland is part of Springer Science+Business Media (www.springer.com)

Foreword

The purpose of the 14th IEEE/ACIS International Conference on Computer and Information Science (ICIS 2015) held on June 28–July 1, 2015 in Las Vegas, USA, was to bring together researchers, scientists, engineers, industry practitioners, and students to discuss, encourage, and exchange new ideas, research results, and experiences on all aspects of Applied Computers & Information Technology, and to discuss the practical challenges encountered along the way and the solutions adopted to solve them. The conference organizers have selected the best 16 papers from those papers accepted for presentation at the conference in order to publish them in this volume. The papers were chosen based on review scores submitted by members of the program committee and consequently underwent further rigorous rounds of review.

In "*SAF*: States Aware Fully Associative FTL for Multitasking Environment", Usman Anwar, Se Jin Kwon, and Tae-Sun Chung propose a new FTL algorithm called SAF. Compared to the previous FTL algorithm, SAF shows higher performance. They also provide performance comparison results of their implementation of SAF and previous algorithm FAST.

In "Security Measures for Web ETL Processes", Salma Dammak, Faiza Ghozzi Jedidi, and Faiez Gargouri present a Common Vulnerability Scoring System (CVSS) and proposes a Meta model for security measure in Web ETL processes enabling security manager to asset anticipated vulnerabilities.

In "Automated Negotiating Agent Based on Evolutionary Stable Strategies", Akiyuki Mori and Takayuki Ito propose a negotiating agent that is based on the expected utility value at the equilibrium point of an evolutionary stable strategy (ESS).

In "Architecture for Intelligent Transportation System Based in a General Traffic Ontology", Susel Fernandez, Takayuki Ito, and Rafik Hadfi introduce an ontology-based system to provide roadside assistance, providing drivers making decisions in different situations, taking into account information on different traffic-related elements such as routes, traffic signs, traffic regulations, and weather elements.

In "Optimization of Cross-Lingual LSI Training Data", John Pozniak and Roger Bradford present a principled approach for making such selection. We present test results for the technique for cross-lingual document similarity comparison. The results demonstrate that, at least for this use case, employment of the technique can have a dramatic beneficial effect on LSI performance.

In "Depth-First Heuristic Search for Software Model Checking", Jun Maeoka, Yoshinori Tanabe, and Fuyuki Ishikawa propose an algorithm called depth-first heuristic search (DFHS), which performs depth-first search but backtracks at states that unlikely lead to an error. Experimental results show that DFHS performs better than current algorithms for both safety and LTL properties of programs in many cases.

In "A Novel Architecture for Learner's Profiles Interoperability", Leila Ghorbel, Corinne Amel Zayani, and Ikram Amous propose a novel interoperable architecture allowing the exchange of the learner's profile information between different adaptive educational cross-systems to provide an access corresponding to the learners' needs.

In "CORE: Continuous Monitoring of Reverse k Nearest Neighbors on Moving Objects in Road Networks", Muhammad Attique, Hyung-Ju Cho, and Tae-Sun Chung present a new safe exit based algorithm for efficiently computing safe exit points of query and data objects for continuous reverse nearest neighbor queries called CORE.

In "A Voice Dialog Editor Based on Finite State Transducer Using Composite State for Tablet Devices", Keitaro Wakabayashi, Daisuke Yamamoto, and Naohisa Takahashi propose a method of editing voice interaction contents using composite state. The results of experiments conducted indicate that this objective was achieved.

In "Analysis of Driving Behaviors Based on GMM by Using Driving Simulator with Navigation Plugin", Naoto Mukai aims for modeling driving behaviors to support operation of novice drivers. Moreover, they examine the effects of navigation at the roundabout intersection for the novice drivers.

In "Bin-Based Estimation of the Amount of Effort for Embedded Software Development Projects with Support Vector Machines", Kazunori Iwata and Elad Liebman, Peter Stone, Toyoshiro Nakashima, Yoshiyuki Anan, and Naohiro Ishii study a bin-based estimation method of the amount of effort associated with code development. They carry out evaluation experiments to compare the accuracy of the proposed SVM models with that of the e-SVR using Welch's t-test and effect sizes.

In "Applying RoBuSt Method for Robustness Testing of the Non-interference Property", Maha Naceur and Lilia Sfaxi propose to apply an approach they developed in a previous work to test the robustness of a very restrictive and important security property, which is non-interference.

In "An Improved Multi-SOM Algorithm for Determining the Optimal Number of Clusters", Imèn Khanchouch, Malika Charrad, and Mohamed Limam focus on multi-SOM clustering approach which overcomes the problem of extracting the number of clusters from the SOM map through the use of a clustering validity index.

In "Conformance Testing for Timed Recursive Programs", Hana M'Hemdi, Jacques Julliand, Pierre-Alain Masson, and Riadh Robbana propose a novel method of offline test generation from deterministic TPAIO. This paper is about conformance testing of timed pushdown automata with inputs and outputs (TPAIO), which specify both stack and clock constraints.

In "Instruction Level Loop De-optimization: Loop Rerolling and Software De-pipelining", Erh-Wen Hu, Bogong Su, and Jian Wang report their work on loop de-optimization at instruction level. They demonstrated their approach with a practical working example and carried out experiments on TIC6x, a digital signal processor with a compiler supporting instruction-level parallelism.

In "ETL Design Toward Social Network Opinion Analysis", Afef Walha, Faiza Ghozzi, and Faïez Gargouri propose an ETL design approach integrating user's opinion analysis, expressed on the popular social network Facebook. It consists of the extraction of opinion data on Facebook pages (e.g., comments), its pre-processing, sentiment analysis and classification, and reformatting and loading into the Data WeBhouse (DWB).

It is our sincere hope that this volume provides stimulation and inspiration, and that it will be used as a foundation for works to come.

Shizuoka University, Japan   Naoki Fukuta
June 2015

Contents

Contributors

Ikram Amous MIRACL-ISIMS Sfax University, Sfax, Tunisia

Yoshiyuki Anan Base Division, Omron Software Co., Ltd., Shimogyo-ku, Kyoto, Japan

Usman Anwar Computer Engineering, Ajou University, Suwon, South Korea

Muhammad Attique Department of Computer Engineering, Ajou University, Suwon, South Korea

Roger Bradford Maxim Analytics, Reston, VA, USA

Malika Charrad Department of Computer Science, RIADI lab, University of Gabès, Gabès, Tunisia; MSDMA Team, Cedric, CNAM, Paris, France

Hyung-Ju Cho Department of Software, Kyungpook National University, Gajang-dong, Sangju-si, Gyeongsangbuk-do, South Korea

Tae-Sun Chung Department of Computer Engineering, Ajou University, Suwon, South Korea

Salma Dammak MIRACL-ISIMS Pole technologique de SFAX BP 242-3021, Sakiet Ezzit Sfax, Sfax, Tunisia

Susel Fernandez University of Alcala. Alcalá de Henares, Madrid, Spain

Faïez Gargouri Institute of Computer Science and Multimedia, University of Sfax, Sfax, Tunisia

Leila Ghorbel MIRACL-ISIMS Sfax University, Sfax, Tunisia

Faiza Ghozzi Institute of Computer Science and Multimedia, University of Sfax, Sfax, Tunisia

Rafik Hadfi Nagoya Institute of Technology, Nagoya, Japan

Erh-Wen Hu Department of Computer Science, William Paterson University, Wayne, NJ, USA

Naohiro Ishii Department of Information Science, Aichi Institute of Technology, Yakusa-cho, Toyota, Aichi, Japan

Fuyuki Ishikawa The University of Electro-Communications, Chofu-shi, Tokyo, Japan; National Institute of Informatics, Chiyoda-ku, Tokyo, Japan

Takayuki Ito Nagoya Institute of Technology, Nagoya, Japan

Kazunori Iwata Department of Business Administration, Aichi University, Nakamura-ku, Nagoya, Aichi, Japan; Department of Computer Science, The University of Texas at Austin, Austin, TX, USA

Faiza Ghozzi Jedidi MIRACL-ISIMS Pole technologique de SFAX BP 242-3021, Sakiet Ezzit Sfax, Sfax, Tunisia

Jacques Julliand FEMTO-ST/DISC, University of Franche-Comté, Besançon, France

Imèn Khanchouch Department of Computer Science, University of Tunis, ISG, Tunis, Tunisia

Se Jin Kwon Computer Engineering, Ajou University, Suwon, South Korea

Elad Liebman Department of Computer Science, The University of Texas at Austin, Austin, TX, USA

Mohamed Limam Department of Computer Science, University of Tunis, ISG, Tunis, Tunisia; Department of Statistics, Dhofar University, Salalah, Oman

Jun Maeoka Research and Development Group, Hitachi, Ltd., Yokohama-shi, Kanagawa, Japan; The University of Electro-Communications, Chofu-shi, Tokyo, Japan

Pierre-Alain Masson FEMTO-ST/DISC, University of Franche-Comté, Besançon, France

Hana M'Hemdi FEMTO-ST/DISC, University of Franche-Comté, Besançon, France; LIP2, University of Tunis El Manar, Tunis, Tunisia

Akiyuki Mori Nagoya Institute of Technology, Aichi, Japan

Naoto Mukai Department of Culture-Information Studies, School of Culture-Information Studies, Sugiyama Jogakuen University, Aichi, Japan

Maha Naceur LIP 2 Laboratory, University of Tunis El Manar, Tunis, Tunisia

Toyoshiro Nakashima Department of Culture-Information Studies, Sugiyama Jogakuen University, Chikusa-ku, Nagoya, Aichi, Japan

John Pozniak Leidos, Chantilly, VA, USA

Riadh Robbana LIP2 and INSAT-University of Carthage, Tunis, Tunisia

Lilia Sfaxi LIP 2 Laboratory, University of Tunis El Manar, INSAT, University of Carthage, Tunis, Tunisia

Peter Stone Department of Computer Science, The University of Texas at Austin, Austin, TX, USA

Bogong Su Department of Computer Science, William Paterson University, Wayne, NJ, USA

Naohisa Takahashi Nagoya Institute of Technology, Nagoya-shi, Aichi, Japan

Yoshinori Tanabe Tsurumi University, Yokohama-shi, Kanagawa, Japan

Keitaro Wakabayashi Nagoya Institute of Technology, Nagoya-shi, Aichi, Japan

Afef Walha Multimedia, Information systems and Advanced Computing Laboratory, University of Sfax, Sfax, Tunisia

Jian Wang Mobile Broadband Software Design, Ericsson, Ottawa, ON, Canada

Daisuke Yamamoto Nagoya Institute of Technology, Nagoya-shi, Aichi, Japan

Corinne Amel Zayani MIRACL-ISIMS Sfax University, Sfax, Tunisia

SAF: States Aware Fully Associative FTL for Multitasking Environment

Usman Anwar, Se Jin Kwon and Tae-Sun Chung

Abstract Nowadays, *NAND* flash memory is begin widely used for data storage purposes in all types of digital devices such as in handheld devices like; MP3 players, mobile phones, and digital cameras or in large scale servers. Reasons of so much popularity of flash memory are its characteristics; low power consumption, non-volatility, high performance, shock resistance and portability. However due to some of its hardware characteristics, a software layer called *flash translation layer* (*FTL*) is used between file system and flash memory. We propose a new *FTL* algorithm called *SAF*. Compared to the previous *FTL* algorithm, *SAF* shows higher performance. We also provide performance comparison results of our implementation of *SAF* and previous algorithm *FAST*.

1 Introduction

NAND flash memory is composed of multiple blocks and each block consists of multiple pages. A type of system software called *FTL* (*Flash Translation Layer*) is used between flash device and file system as shown in the Fig. 1. Main functionality of *FTL* is to provide address translation. *FTL* translates the logical addresses issued by the file system to the physical addresses. *FTL* maintains a table which contains the logical to physical location mapping information. *NAND* flash memory also has some drawbacks because of its hardware architecture. It has an `erase-before-write` architecture [1]. That is, if a previously written location is needed to be updated then the location has to be erased first before new data can be written on it. Also pages of a block can only be programmed in sequential order as random ordered programming of pages becomes cause of major performance degradation [2].

U. Anwar · S.J. Kwon · T.-S. Chung (✉)
Computer Engineering, Ajou University, Suwon, South Korea
e-mail: tschung@ajou.ac.kr

U. Anwar
e-mail: usmananwar@outlook.com

S.J. Kwon
e-mail: sejin1109@ajou.ac.kr

© Springer International Publishing Switzerland 2016
R. Lee (ed.), *Computer and Information Science 2015*,
Studies in Computational Intelligence 614, DOI 10.1007/978-3-319-23467-0_1

Fig. 1 Interleaved instructions in multitasking environment

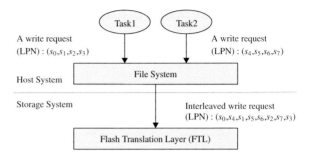

Although flash memory has replaced the conventional disk storage devices but still due to its aforementioned hardware characteristics its performance is not up to the mark in some specific computing environments, one of them is multitasking environment. Nowadays almost all general purpose operating systems (Windows, Linux, Mac OS) support multitasking. In multitasking environment multiple applications perform storage access operations simultaneously and to address the storage access requests from all applications, file system interleaves the storage access requests issued by all applications and pass it to the storage device [3].

As shown in Fig. 1, file system interleaves the write requests from different tasks and passes to the *FTL* and it will consider these write requests as random requests even though these requests were originally issued in sequential order by different tasks, this affects the performance of flash memory badly.

In this study, we propose a new *FTL* algorithm for *NAND* flash memory which avoids the performance degradation of flash memory caused by costly merge operations performed to handle random write requests. It not only just ensures the minimum cost of each merge operation but also reduces the frequency of merge operations which results in lower number of expensive erase operation. Simulation results of our proposed algorithm shows much better performance comparing to famous previous *FTL* algorithm *FAST* [4].

2 Background

In this section, we will describe the address mapping techniques which are being used in flash memory and different types of merge operation and also the cost comparison of these merge operations.

2.1 Address Mapping Techniques

Address mapping techniques can be classified into three types; sector-mapping, block-mapping and hybrid-mapping. In sector mapping, each logical sector is mapped to a physical sector. Sector mapping technique requires a lot of *DRAM* space to maintain the table thatch why it is not suitable for resource-limited flash memories [5].

To overcome the excess use of memory by earlier mentioned technique, block mapping technique was proposed. Block mapping scheme maintains physical blocks table instead of sectors, which requires less memory in *DRAM* to store the table. However, if the file system issues writes commands with the same logical page number (LPN) many times then *FTL* will require to perform merge operation so many times which leads to severe performance degradation. Since both page and block mapping have some drawbacks as mentioned above.

A hybrid mapping scheme was introduced which is combination of previously defined mapping schemes. It first gets the corresponding physical block number by using block mapping scheme then gets the empty page within the physical block by using page-mapping. Hybrid mapping is widely used in log buffer based *FTL* algorithms. In this type of *FTL*, flash memory blocks are divided into two types; data blocks and log blocks. Generally a data block is associated to a log block. When the data block gets full then the new updates to this data block will be redirected to its associated log block to delay the merge operation and old data in data block becomes invalid data [2].

2.2 Merge Types and Cost Comparison

In log buffer based algorithms, *FTL* reclaims the memory blocks with invalid data, by using different types of merge operation. There are three types of merge operations; switch, partial and full merge operation. When a log block is filled in sequence (in-place order) then switch or partial merge operations are carried out.

As shown in Fig. 2a, since log block contains all valid, sequentially written pages corresponding to data block, a simple switch operation is performed whereby log block becomes new data block and the old data block is erased. Figure 2b illustrates partial merge operation between data block and log block where only the valid pages in data block are copied to log block and the original data block is erased changing the log block to new data block.

Full merge involves the largest overhead among the three types of merges. As shown in Fig. 2c, the valid pages from the log block and its corresponding data block are copied into a new data block and old data and log blocks are erased. As we can see that for switch operation we do not need to copy any page rather we just covert the log block to new data block, thats why it is the cheapest merge operation. In this

Fig. 2 Merge operations.
a Switch operation. **b** Partial
merge operation. **c** Full
merge operation

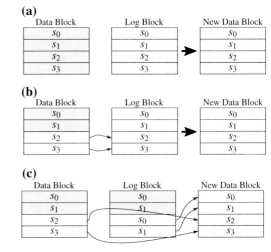

paper, to analyze the cost of partial merge operation and full merge operation, we
will use following variables;

- n_a: number of associated data blocks to a log block
- v_pages: number of valid pages to be copied
- t_c: time required to copy a page
- t_e: time required to erase a block

Following is the equation acquired by using above variables, for partial merge operation cost;

$$partial\ merge\ cost = t_c * v\_pages + t_e \qquad (1)$$

Partial merge operation cost includes the expenses of copying the valid pages from
old data blocks to the new data block and erasing the old data block. Whereas, full
merge operation is much more expensive than the partial merge as show below;

$$full\ merge\ cost = n_a(t_c * v\_pages + t_e) + t_e \qquad (2)$$

We can see that the cost deeply depends on the number of associated data blocks to
a log block n_a. As n_a increases, the cost of full merge operation also increases.

2.3 Previous Work

There have been much research done on *FTL* and many well known algorithms have
been proposed, very famous one of them is fully associative page translation (*FAST*)
which is based on the log buffer-based scheme. *FAST* improves the performance

of previously proposed algorithm *BAST* [4]. In *BAST* a single log block can be associated to only one data block and random update patterns can easily degrade the performance of flash memory because it does not allow the log blocks to be shared among different data blocks since *BAST* has 1:1 association between data and log block. Therefore, *BAST* is often forced to reclaim a log block which has not been fully utilized. This phenomenon is called block threshing problem.

Whereas in *FAST* scheme, a log block can be associated to multiple number of data blocks. *FAST* scheme was proposed to solve the problem of *BAST*. *FAST* always utilizes the log block capacity till the end before allocating another log block. This improves the problem of block threshing but delays the process of reclaiming a single log block because in *FAST* a log block can be associated to data blocks of as many of the number of pages in a block. For *FAST*, the maximum value of n_a can be equal to the number of pages in a log block. For sequential writes *FAST* only allocates the single SW log block and this will be issued to each update request having offset zero. This also generates the block threshing problem in SW log block. If the sequential writes are interleaved by the file system then the *FAST* will perform poorly because this will not only increase the number of partial merge operation because of block threshing, but also increases the number of full merge operation. In this paper we will focus on the issues described above and provide a solution for them in the next section.

3 SAF

SAF is based on the log buffer based technique. Flash memory is divided into two types of blocks; data blocks and log blocks. Log blocks are further divided into two subtypes; Sequential write (SW) and Random write (RW) log blocks. Unlike *FAST* scheme, *SAF* allocates multiple SW blocks instead of only one, to decrease the merge cost by increasing the number of switch merge operations as we have seen in the previous section that switch operation is the optimal of all three merge operations. *SAF* uses two page mapping table, for SW and RW log blocks, like *FAST*. As we have described above that in multitasking environment, file system issues storage access commands from multiple applications in interleaving fashion which causes the flash memory degradation. For this, compared to previous work, we introduce states for log blocks of flash memory. We use two different sets of states to handle sequential and random write requests. This state information can be stored within the page mapping table of SW and RW log blocks.

3.1 Handling Sequential Requests

As described above that *FAST* algorithm does not handle sequential write request efficiently so we propose following states for SW log blocks to extract the sequential streams from the interleaved I/O instructions issued by the file system to increase the probability of cost effective switch operation. Each SW log block is in one of the state at some point.

- *F* (Free) state: If a sequential log block has not been written or has been erased, the block is said to be in *F* state.
- *A* (Allocated) state: If a sequential log block has been associated to a data block then it is said to be in *A* state. A sequential log block can be in *A* state for a certain period of time which we call allocation time. If sequential log block with state *A* gets some further update within the allocation time then time will be increased by one unit but if its allocation time gets over then its state will be changed to state *M*.
- *M* (Ready for merge) state: A sequential log block with state *M* means that this block is ready for switch merge operation either because it got full or it did not get any update request within its allocation time.

SAF will consider each update request with offset equal to zero for the SW log block but it will get the SW log block only if there is any SW log block available otherwise it will be handled as random write which is described in next section. Initially all the SW log blocks are given *F* state. When *SAF* receives the update request for a data block, with the offset zero, it will first try to find any SW log block with the state *F*. If it does not get any SW log block with *F* state then it will get the SW log block by merging any previously allocated SW log block with state *M*. If there is also no SW log block available with state *M* then the update request will be redirected to random log blocks and will be handled as random request. But if it finds the SW log block, data will be written into it and SW page mapping table will be updated also the SW log block state will be changed to *A*. SW log block, with state *A* will be allocated to particular data block for some predefined allocation time. If SW log block with state *A* gets some further update within the allocation time then it will get increased by one but if its allocation time gets over then its state will be changed to state *M* which means that this SW log block is ready to get merged.

This scheme will not only let the SW log blocks to get full even if the file system is not issuing the commands in sequential order but also addresses the simultaneous I/O streams from the file system because of the utilizing the multiple sequential log blocks rather than just using single block and making it candidate for each update request with offset equal to zero.

Suppose we have 3 data blocks (DB1, DB2, DB3) and 2 SW log blocks (SW1, SW2) and each block contains 4 pages as shown in the Fig. 3. When updates on data blocks are issued by file system, *SAF FTL* redirects them to log blocks by invalidating the corresponding pages in the data block which are shown by gray color. For instance, assume that the update pattern sequence is $s_0, s_5, s_8, s_1, s_9, s_2, s_6, s_{10}, s_3, s_7, s_{11}$. *SAF* will allocate the SW1 to DB1 for first write request s_0 and also SW1 will be allotted a specific time as we can see in the time line for SW1. SW1 is now in state *A*. For further instructions from given pattern; s_5 be handled by the RW log blocks because it does not have offset equal to zero but the allocation time for SW1 will be decreased by one unit as shown in the time line. For s_8 *SAF* will assign SW2 to

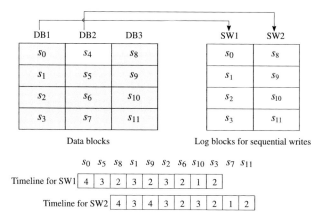

Fig. 3 Handling sequential requests

data block DB3 for some time and its state will be changed to state A. s_1 will be addressed by the SW1 and one unit of time will be added to its allocation time. s_9 will be addressed by the SW2 and s_6 will be handled by the RW log blocks one unit of time will be reduced from SW1 and SW2 allocation time. Once a SW log block run out of time allocation time or gets full then its state will be changed to M which means its ready to get merged.

3.2 Handling Random Requests

We also define following states for RW log blocks to handle the random write requests and each RW log block is in one of the state at some point. In the following definition, *associated data block* means a data block which has been associated to a respective random log block and *non associated data block* means a data block which has not been associated to a respective random log block.

- E (Empty) state: If a random block has not been written then it is said to be in E state.
- O (Open) state: A random block in this state can accept update request from both, associated data block and non associated data block.
- B (Blocked) state: A block with state B means that it can accept update request only from associated data block.

SAF fills all the RW log blocks simultaneously rather than conventional sequential way in which one RW log block gets associated to a data block only if the already associated RW log block gets full. *SAF* assigns the above mentioned states to RW log blocks, to ensure that each RW block gets associated to even number of data blocks. Initially, all the RW blocks are in state E. When an update request is addressed by a RW log block with state E or O then its state is changed to state B, which means that this

s_0	s_4	s_8	s_{12}	s_{16}		s_1	s_{14}	s_{19}
s_1	s_5	s_9	s_{13}	s_{17}		s_2	s_{11}	
s_2	s_6	s_{10}	s_{14}	s_{18}				
s_3	s_7	s_{11}	s_{15}	s_{19}				
DB1	DB2	DB3	DB4	DB5		RW1	RW2	RW3
		Data blocks					Log blocks for random writes	

I/O pattern	RW1	RW2	RW3
	E	E	E
s_1	B	E	E
s_{14}	B	B	E
s_2	B	B	E
s_{19}	O	B	B
s_{11}	B	O	B
s_5	B	B	O

Fig. 4 Handling random requests

RW log block is no longer available to address update request from non-associated data blocks, and the state of the very next RW block will be changed to state O if its previous state was B. This process continues in circular manner which ensures that each log block gets associated to as less number of data blocks as possible, in other words its trying to minimize the variable n_a as described above to reduce the full merge cost.

This scheme works very efficient if we use multiple RW log blocks. *SAF* also does not wait for all the RW log blocks to get full, for a merge operation. For example if there are three RW log blocks and one gets full then *SAF* will perform the full merge operation to reclaim this RW log block instantly so that further update request can be divided among all three RW log blocks rather than just two RW log blocks. In *KAST* [6], a RW log block is reclaimed once all other random log blocks got full which causes the further update requests to be divided among lesser number of random log blocks which in process causes a RW log block to get associated to more number of data blocks.

Figure 4 showing the process of handling random write operation. We have five data blocks and three RW log blocks. Assume we have following I/O pattern $s_1, s_{14}, s_2, s_{19}, s_{11}, s_5$. *SAF* will associate the RW1 to DB1 for first write request s_1 and its state will be changed to B from E. s_{14} will be entertained by the RW2 and its state will be changed to B. s_2 will be handled by the RW1 because it is already associated to DB1 even though its in the B state. s_{19} will be addressed by the RW3 and its state will be converted to B and because this is the last RW log block thats why the state of first RW log block will also be changed to state O. Same procedure will be repeated for further instructions.

4 Experiments

We evaluated the performance of *SAF* using simulation. We used I/O trace of three different applications accessing the storage device simultaneously on windows based system. This I/O trace contains sequential and random storage access instructions from all three applications which are interleaved by the file system. We used our own simulator which simulates the working of *NAND* flash memory consisting of 8192 blocks, each block containing 32 pages. We used same number of total log blocks for *SAF* and *FAST*. For *SAF*, we used four log blocks for random write operation and three log blocks for sequential write operations and for *FAST*, we used 6 log blocks for random write operations and 1 for sequential write operation as described in [4].

Figure 5 shows the comparison of *SAF* and *FAST* simulation results. We can see that *SAF* scheme has clearly outperformed the *FAST* in terms of total number of merge operations even both schemes are using same number of log blocks. This is because *SAF* tries to increase the number of switch operation and partial merge operation by not only using the multiple sequential log blocks but also by extracting the sequential I/O streams from interleaved instructions from the file system by using allocation time for each sequential log block. Because of the smaller number of merge operation *SAF* has also decreased the number of erase operation as we can see in Fig. 6 which is very effective.

SAF also ensures that each random log blocks get associated to as minimum number of data block as possible. Figure 7 showing the average number of associativity (n_a) of a log block to data blocks, for both schemes. We can see that *SAF* is able to keep the random log block associativity n_a value smaller than the *FAST* scheme which in result decreases the cost of full merge operation.

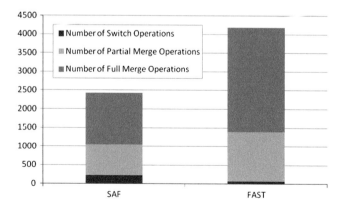

Fig. 5 Merge operation counts

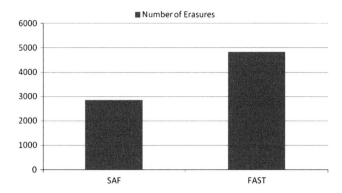

Fig. 6 Number of erasures

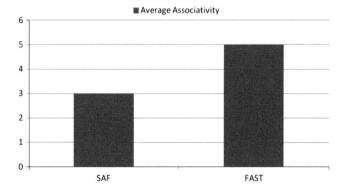

Fig. 7 Average associativity

5 Conclusion

In this paper, we proposed a new *FTL* algorithm called *SAF*. The main idea of *SAF* is to minimize the total number of merge operations and decrease the cost of full merge operation by introducing the states for log blocks which in case of sequential log blocks, maximizes the probability of switch merge operation by delaying the partial merge operation of SW log block and for random log blocks, tries to get them associated to minimum number of data blocks. Compared to previous work, our experiments shows that *SAF* performs well in multitasking environment than *FAST*.

Acknowledgments This research was supported by Basic Science Research Program through the National Research Foundation of Korea (NRF) funded by the Ministry of Education (2013R1A1A2A10012956, 2013R1A1A2061390).

References

1. Chung, T.-S., Park, D.-J., Park, S., Lee, D.-H., Lee, S.-W., Song, H.-J.: A survey of flash translation layer. J. Syst. Archit. **55**(5–6), 332–343 (2009). doi:10.1016/j.sysarc.2009.03.005
2. Ma, D., Feng, J., Li, G.: A survey of address translation technologies for flash memories. ACM Comput. Surv. **46**(3), 36:1–36:39, (2014). doi:10.1145/2512961
3. Lebre, A., Huard, G., Denneulin, Y., Sowa, P.: I/o scheduling service for multi-application clusters. In: 2006 IEEE International Conference on Cluster Computing, September 2006, pp. 1–10
4. Lee, S.-W., Park, D.-J., Chung, T.-S., Lee, D.-H., Park, S., Song, H.-J.: A log buffer-based flash translation layer using fully-associative sector translation. ACM Trans. Embed. Comput. Syst. **6**(3) (2007). doi:10.1145/1275986.1275990
5. Chung, T.-S., Park, H.-S.: Staff: a flash driver algorithm minimizing block erasures. J. Syst. Archit. **53**(12), 889–901 (2007). doi:10.1016/j.sysarc.2007.02.005
6. Cho, H., Shin, D., Eom, Y.I.: Kast: K-associative sector translation for nand flash memory in real-time systems. In: Design, Automation Test in Europe Conference Exhibition, DATE '09, April 2009, pp. 507–512

Security Measures for Web ETL Processes

Salma Dammak, Faiza Ghozzi Jedidi and Faiez Gargouri

Abstract Security aspects currently play a vital role in software systems. As security managers have to operate within limited budgets they also have to patch up the increasing number of software security vulnerabilities. They need to perform a risk evaluation in order to determine the priority of patching-up vulnerabilities. The use of quantitative security assessment methods enables efficient prioritization of security efforts and investments to mitigate the discovered vulnerabilities and thus an opportunity to lower expected losses. Elsewhere, Extraction Transformation Load (ETL) processes, known as a core, development of WeBhouse. Securing these processes is highly important and helps in mitigating security defects in decisional system. For this purposes, this paper adopts the Common Vulnerability Scoring System (CVSS) and proposes a Meta model for security measure in Web ETL processes enabling security manager to asset anticipated vulnerabilities.

1 Introduction

WeBhouse represents a new form of data warehouse that collects data from the web, which presents a large source of information, in order to integrate the data from several sources. Due to sensitive data contained in the DW, it is important to assure the security of the decisional system from the early stages of the life cycle of its development. In the Web context, the mega-volumes of data, the increasing number of users and the heterogeneity of requirements increase new challenges, as the nature of data, the lack of control over the sources, and the frequency of changes on data.

S. Dammak (✉) · F.G. Jedidi · F. Gargouri
MIRACL-ISIMS Pole technologique de SFAX BP 242-3021,
Sakiet Ezzit Sfax, Sfax, Tunisia
e-mail: damak.salma@gmail.com

F.G. Jedidi
e-mail: jedidi.faiza@gmail.com

F. Gargouri
e-mail: faiez.gargouri@gmail.com

© Springer International Publishing Switzerland 2016
R. Lee (ed.), *Computer and Information Science 2015*,
Studies in Computational Intelligence 614, DOI 10.1007/978-3-319-23467-0_2

13

Besides, the main challenge in this field is how to integrate heterogeneous and safety data in the WeBhouse when some or most data sources reside on the Web.

Indeed, the use of Internet in a storage process (acquisition, storage and access) increases the risk of attacks that can affect decisional system and make it more vulnerable. An improper disclosure of such data represents a risk to the system. These drawbacks lead us to consider that security is a key factor for a WeBhouse. In addition, the security needs are not considered at the earliest designing phases of data warehouses; also most design methods do not integrate them. An engineer who does not have a complete description about security requirements and the way of securing data will treat these aspects at the closing stages of the WeBhouse development.

Note that the Extraction-transformation-loading (ETL) processes play an important role in WeBhouse architecture because they are responsible for integrating data from heterogeneous data sources. But, works dealing with ETL processes security are few and far from proposing a complete approach to secure the ETL web process. They are focused mostly on the extraction process although the transformation process covers major complex operations. Also, the loading process requires a protected network to charge data in the WeBhouse. It is therefore widely recognized that the appropriate design and security of the ETL processes are key factors in the success of WeBhouse projects.

At the design and implementation stages of WebHouse life cycle, vulnerabilities are inadvertently introduced and result from defects or weaknesses. They can be exploited by attackers to harm the system. Therefore, security needs to be considered and measured from the early stage of the development life cycle. To evaluate the security and safety level in WEB ETL processes, we define a secure Meta model. Our meta enriches the Meta-model of UML activity diagram defined by OMG [1] by adding security classes based on the metric groups of the CVSS (Common Vulnerability Security Score) [2]. Our approach enables us to verify the safety of the data source before the extraction process, to detect vulnerability and define for each one the suitable preventions and solutions and to secure the inter-phase transition. The proposed solutions integrating security needs are melted in the CVSS factor metric groups values, thus enabling to asset the severity impact of vulnerabilities. This measuring allows us to prioritize vulnerability and reduces the impact of vulnerabilities on the ETL process because the most severe will be the first remedy. Our proposition is applied to a case study for the Web ETL processes of an online banking system, due of the risk that may come across this system and the sensitivity of the data that they process.

This paper is organized as follows: Sect. 2 surveys ETL process security and security measurement related works. Section 3 presents our secure Web-ETL Meta-model integrating vulnerabilities specification. Section 4 depicts our secure web-ETL case study instantiating our meta model in banking WeBhouse ETL processes. Section 5 concludes the paper with some pointers to future directions.

2 Related Work

2.1 ETL Process and Security

2.1.1 Multidimensional Integration of Web Data

The proposed Web Warehouse architecture in [3] is composed of two parts: a repository of functional characteristics of a data warehouse and a repository of Web click-streams describing the data. Before data extraction, the approach requires the verification of the accuracy of data sources from the Web. Web logs are unified in [4] the proposed approach is to represent the raw data log file in a model that conforms to a meta-model of generic Web log. Once this model is designed, a multidimensional conceptual model is obtained automatically through the transformation language QVT (Query/View/Transformation). Based on the SOA architecture, [5] discusses the real-time data analysis and they design the real time data warehouse architecture. In the proposed architecture, a multilevel real-time data cache is introduced to facilitate the realtime data storage. Reference [5] combines the Web services and data warehouse by using the Web service to capture real-time data and make the real-time data capture and load easily.

However, to summarize the current literature, we note that all the work deal with Web inputs. We find that the treatment of ETL processes which present an important task for designing the WeBhouse is infrequent or absent. In addition, non-functional security requirements are not addressed in any design approaches with the exception of work [6] which has emphasized the need to verify the accuracy of data sources. Therefore, in the next section, we present the few works studying the safety of ETL processes.

2.1.2 ETL and Security

Although there are conceptual proposals for the modeling of ETL processes, there are few proposals that support safety issues. The complexity of the ETL system arises as the sources use different data formats, which has to be cleaned, transformed, aggregated and loaded into the data warehouse as homogeneous data. Consequently and with the emergence of the web, different security problems are triggered and a security approach for the ETL process becomes a urgent need. Reference [7] proposes a simulation model for the extraction of secure data during the ETL process. To accomplish the conceptual modeling of Secure Data Extraction in ETL processes, they extend the UML uses case and sequence diagrams. An algorithm is proposed based on the users authentication, the encryption of the extracted data, the verification of data corruption. This work is extended in [8] where authors analyze the ETL process under two security metrics: vulnerability index and security index. A framework for securing any phase in the ETL process has been suggested together with a methodology for assessing the security of the system in the early stages. Reference

[8] develops a simulation tool SeQuanT which quantifies the security of the system, in a general context. An object-oriented approach to model the ETL process embedding privacy preservation is proposed in [9] the researchers present a class diagram for ETL embedding Privacy Preservation. The main components of this diagram are Data Source, retrieval component, source identifier, join, PPA and DSA. The PPA is the most important and major component of this architecture, its task includes cleaning and transforming data followed by privacy preserving attributes.

There are few research papers on ETL security. These researches do not take into account non-functional security needs of ETL processes in the Web business level and they do not address the vulnerabilities between processes of Web ETL.

2.2 Security Measurement

To secure a system, we must identify and assess vulnerabilities. So we need to prioritize these vulnerabilities and remediate those that pose the greatest risk. Several research papers addressing topics of measuring information security have been written in the last few years. National Institute of Standards and Technology have published Special Publications dealing with topics of Risk Management and Information Security Metrics [10–12].

Several studies are based on the CVSS which is an adopted standard that enables security analysts to assign numerical scores to vulnerabilities according to severity to measure security [13]. In [14], authors propose the novel approaches of handling dependency relationships at the base metric level, and aggregate CVSS metrics from different aspects. A CVSS risk estimation model is presented in [15]. The proposed model estimates a security risk level from vulnerability information as a combination of frequency and impact estimates derived from the CVSS.

3 The Meta Model Proposed for Secure WEB ETL

To model our secure Web ETL processes, we extend the Web ETL meta model presented in [16], based on the activity diagram Meta model UML OMG [1] and we enrich it with our security Meta classes. Because we cannot improve what we cannot measure, [17, 18], measuring security in Web ETL process is an essential. By applying security metrics, we can judge the relative effectiveness of our proposal [13]. The Common Vulnerability Scoring System (CVSS) is an adopted standard that enables security analysts to assign numerical scores to vulnerabilities according to severity [13]. It calculates the vulnerability score using attributes grouped into three metric groups: base, temporal and environmental. A CVSS score is a decimal number in the range (0.0, 10.0), where the value 0.0 has no rating (there is no possibility to exploit vulnerability) and the value 10.0 has full score (vulnerability easy to exploit). The base metric group (mandatory) quantifies the intrinsic characteristics

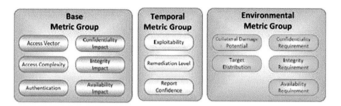

Fig. 1 CVSS metrics groups

of vulnerability in terms of two sub-scores: (i) exploitability subscore; composed of access vector (AV), access complexity (AC) and authentication instances (Au), and (ii) impact subscore to confidentiality (C), integrity (I) and availability (A) [14] (Fig. 1).

Scoring equations and algorithms for the base, temporal and environmental metric groups are described below. The base equation is the foundation of CVSS scoring. The base equation is:

$$BaseScore = ((0.6 \times Impact) + (0.4 \times E)1.5) * f(Impact) \tag{1}$$

With:

$$Impact = 10.41 \times (1 - (1 - C) \times (1 - I) \times (1 - A))$$

$$Exploitability = 20 \times AV \times AC \times Au$$

$$f(impact) \begin{cases} = 0 & if\ impact = 0 \\ = 1.176 & otherwise \end{cases}$$

The temporal (optional) equation will combine the temporal metrics with the base score to produce a temporal score ranging from 0 to 10. Further, the temporal score will produce a temporal score not higher than the base score and it is composed of Exploitability (E), Remediation Level (RL) and Report Confidence (RC). The temporal equation is:

$$TemporalScore = (BaseScore \times E \times RL \times RC) \tag{2}$$

The environmental (optional) equation will combine the environmental metrics with the temporal score to produce an environmental score ranging from 0 to 10. The environmental equation is with CDP (CollateralDamagePotential), TD (TargetDistribution), (CR/IR/AR) security requirement respective to Confidentiality/Integrity/Availability.

$$EnvironmentalScore = ((AT + (10 - AT) \times CDP) * TD) \tag{3}$$

With:

$$AdjustedTempora(AT) = TemporalScore \; recomputed \; with \; the \; Base \; Scores$$
$$Impact \; subequation \; replaced \; with \; the \; AdjustedImpact \; equation$$

$$AdjustedImpact = min(10, 10.41 \times (1 - (1 - C \times CR) \times (1 - I \times IR)$$
$$\times (1 - A \times AR)))$$

In our Meta model, we represent the Environmental Metric Group in a Meta Class while the Temporal Metric Group factors are represented in the vulnerability catalog. Some of the Base Metric Group factors are presented as Meta class like Access, Security Aspect, others are presented as Meta class association like the Authentication, the Access Complexity.

In order to consolidate the anticipated vulnerabilities in the WEB ETL process, we define a Metaclass VulnerabilityCatalog. This Meta class contains the most critical vulnerability that can be detected during the three WEB ETL processes (Extraction, Transformation, Loading) and we take into account the vulnerabilities encountered between processes. The meta class Vulnerability inherits all the Vulnerability Catalog attributes and has the state attribute having values: detected, corrected, expected. This categorization minimizes the probability that vulnerability happens by proposing preventive actions and enables to define a secure environment. Also, it accelerates the remediation actions since the security manager knows from the beginning the most of the vulnerabilities that can be detected and he has already prepared the appropriate remediation solutions.

As mentioned previously, our Meta model is composed of two packages: the Web ETL Meta model and the security Meta model. In the following part, we describe the Web ETL Meta model represented in Fig. 2:

- Steps inherit from "ActivityGroup" class. It models the extraction, transformation and loading processes. Each "Step" is composed of "Activities".
- Sons and Parents inherit respectively from the "inputPin" and the "outputPin" classes.
- Xml Buffer, Log file, Website and Relational DB inherit from the DataStoreNode class are used as storage means.
- XML Dimensional is an XMLBuffer class. It represents a dimensional structure (fact, dimension, attributes ...).

The security Meta Model is presented with blue color in Fig. 3 and described as follow:

- Security Need represent user requirements that must be satisfied in our security approach.

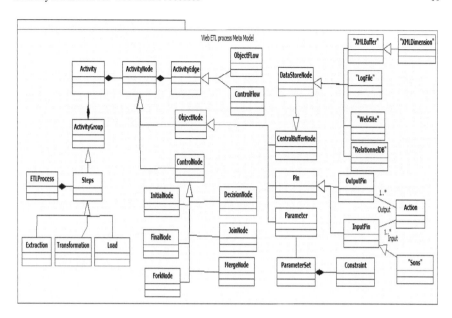

Fig. 2 Web ETL meta model

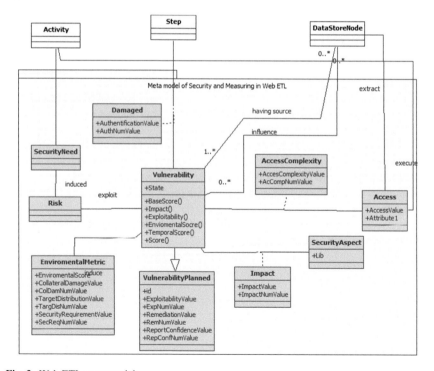

Fig. 3 Web ETL meta model

- Risk is security problems preventing the achievement of security goals.
- Vulnerability Catalog is the anticipated vulnerability having as attribute the three factors of the temporal Metric.
- Group: confirmation of the technical details of vulnerability, the remediation status of vulnerability, and the availability of exploiting code or techniques. Since temporal metrics are optional, they include a metric value that has no effect on the score. This value is used when the user feels that the particular metric does not apply and wishes to skip over it.
- inherits from the Vulnerability Catalog Meta class.
- Access present the type of access to the data (local, adjacent network).
- Security Aspect are Confidentiality, Integrity, Availability.
- Environmental Metric Group captures the characteristics of vulnerability that are associated with a users IT environment.
- Impact is an association class between the Security Aspect and the Vulnerability and measures the impact value.
- Affect is an association class between the Access and the Vulnerability and measures the Access Complexity.
- Damage is an association class between the Steps and the Vulnerability and presents the authentication metric.

4 Scoring Security Measure in Banking WeBhouse ETL Processes

Our proposition is applied to secure the Web ETL processes of a banking system. We focus on each process activity and define for it vulnerabilities that can affect quality and safety of data. To achieve this purpose, our Meta model, covering ETL process and CVSS vulnerabilities, is instancied. The first step in our work is the definition of non-functional requirements depicted by SecurityNeed. Next, we search and set, for the defined requirement the major risks that threaten the meeting of these needs. This work leads us to set vulnerabilities and categorize them. So, for each vulnerability we identify the affected activity during the process (Extract, transform, load), the access type, the security aspect that it affects, the environmental and temporal factor. All these parameters enable us to calculate the CVSS score and to set the safe remediation solution.

4.1 Extraction Process

At the Extraction phase, some specific activities are defined to integrating several web sources, especially log files and websites [16]. This step consists of extracting data from log files (stored in web servers), websites and sources (relational DB, XML ...).

Table 1 Security risks in the WEB extraction process

Risk	Vulnerability
Improper input	Violation of customer private information
	Erroneous and incorrect URL data
	Entrusted Website
	Entrusted data (virus, malware)
	Lack of integrity controls
	Missing encryption of sensitive input
Administrator identity problem	Unprotected transport credentials
	Weak cryptography of password
	Inssufficient verification of data authenticity
	Improper autorisation
	Lack of access control
	Cybercrime: Phishing

The main objective of this step is to unify these sources, into a single format to be used later. Our task consists in controlling the safety of source and to secure the extracted data. In Table 1 presents the majority of security problems intended in the Extraction phase. Two risks are defined and for each one the corresponding vulnerabilities are specified.

In the next paragraph, we will give an example of how to present security in the activity diagram 'Log file structuring'. Structuring log file is an essential activity that allows structuring and organizing a log file in an XML form. We study two vulnerabilities which can be detected in this step and calculate the CVSS score.

Unprotected Transport of Credentials

Figure 4 presents 'LogFileStructure' activity describing the log files structuring process enriched by the security problem that can decrease the process quality. To measure the vulnerability Unprotected transport of credentials, we set values for each factor of the three metric groups of CVSS: base metric, environmental metric and temporal metric. For the base metric, this vulnerability affects greatly (AC: 0.35) the access by an AdjacentNetwork (AT: 0.646).). It does not damage the authenticity (Au: 0). It affects completely Confidentiality (C: 0.660), and has no impact on Integrity (I: 0) and availability (A: 0).

As an environmental factor, the detected vulnerability increases the collateral damage (CDP: 0.5) and the target distribution (TD: 1). It affects security requirements by increasing confidentiality and reducing integrity and availability ([CR/IR/AR], [1.51/0.5/0.5]). For the temporal factor, this vulnerability has a Proof of concept Exploitation value (E: 0.99) i.e. the code or technique is not functional in all situations and may require substantial modification by a skilled attacker. It has a Workaround Remediation Level (RL: 0.95) i.e. there is an unofficial solution available and it has an Uncorroborated Report Confidence (RE: 0.95). Based on the defined factors value in Fig. 4, we calculate the Base, Temporal and Environmental score.

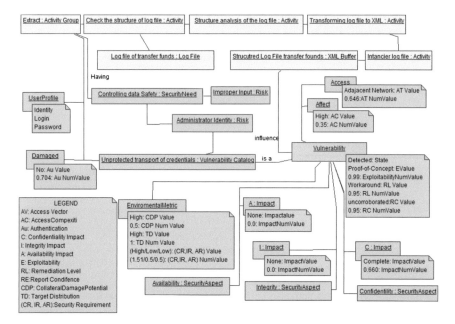

Fig. 4 'Unprotected transport of credentials' security factors in the "Log File Structure" activity diagram

$$BaseScore = 4,58$$
$$Temporal\ Score = 3,72$$
$$Environmental\ Score = 2,93$$

Violation of Customer Private Information

In Fig. 5, we present the same activity but the security need protection of the authentication parameters is treated. The CVSS score are calculated based on the corresponding CVSS factors value. As shown in Fig. 5, we have specified the respective factor for the vulnerability 'Violation of customer private information' and set values for each factor of the three metric groups of CVSS. We note that this vulnerability affects a lot (AC: 0.35) the Network (AT: 1). It has no impact in Confidentiality (C: 0), but affects completely Integrity (I: 0.660) and partially Availability (A: 0.275). It keeps the same values as the previous vulnerability at the temporal factors. At the environmental factor, we notice that this vulnerability does not damage system (CDP: 0) and has a medium distribution (TD: 0.75). It affects security requirements by augmenting integrity completely and availability partially and has no impact in confidentiality ([CR, IR, AR], [0.5/1.5/1]). With these values, we will calculate the Base, Temporal and Environmental score.

$$BaseScore = 5,89$$
$$Temporal\ Score = 5,27$$
$$Environmental\ Score = 3,87$$

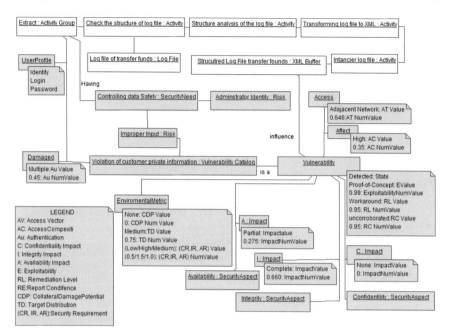

Fig. 5 'Violation of customer private information' security factors in the "Log File Structure" activity diagram

Knowing that the base metric group is the mandatory score in CVSS, we compare the two calculated results. The base score of Violation customer private information is greater than the base score of 'Unprotected transport credentials'. This observation leads us to consider that the first vulnerability is more severe and it impact highly the process. It should be treated quickly and strict and relevant security solutions should be proposed to avoid any attempt to attack private information of bank's customer.

4.2 Transformation Process

In ETL design process, the transformation step is an important and complex task. There are a set of web transformation activities specific to web sources. These activities take as input web data sources structured in the Extraction process. At this level, the data should be protected from intrusion and undesirable access. These risks affect the data quality and safety. In Table 2, we define and categorize the anticipated vulnerability of the transformation process.

Table 2 Security risks in the WEB transformation process

Risk	Vulnerability
Network problem	Lack of access control between process
	Risk of loss data
	Problem of downtime during the treatment and/or the transfer of data
	Data corruption during the transmission to the Loading process
	Risk of losing business data (Customer Private data, Accounts number,...)
Data safety	Joining erroneous data with incorrect values or misspelled attribute
	Missing encryption of accounts information
	Missing encryption of customer private information
	Improper autorisation
	Transmission of sensitive data to the loading process without encryption
	Erroneous business web data
	Missing encryption of sensitive data
	Cybercrime: Phishing

This table represents a catalog helping the security manager to find the anticipated vulnerabilities and to define for each one the corresponding security remediation. This goal is achieved by relying the CVSS factors value prioritizing vulnerability and facilitating the choice of the security solution.

4.3 Load Process

Finally, a secure environment should be available to the loading process which loads the processed data to the WeBhouse. In this stage, many problems are detected and should be remediated to assure the WeBhouse data safety. In Table 3, we list some security problems which can be detected in the loading process.

Table 3 Security risks in the WEB Loading process

Risk	Vulnerability
Data safety	Risk of loss data
	Downtime problem during the data loading
	Unprotected DSA

5 Conclusion

An early detection of security requirement has an impact on the further design decisions providing better security specifications and final products. Security constraints should be considered in the whole development process from early stages to final tools. Because the ETL processes present the first stage in the WebHouse development and it treats an important volume of data coming from heterogeneous and uncontrolled sources, our challenge is to secure the WEB ETL processes and offer a safe environment to modulate the extracted data. In this paper, we propose an approach to secure Web ETL processes by describing the anticipated vulnerabilities for each step and measuring the Score of each one based on the CVSS factor. This measuring prioritizes the critical vulnerabilities and helps the security manager to define the suitable solutions. Our future work consists in the implementing of our proposal and defining a tool helping designer from meta to model while treating non-functional security requirements from the business level.

References

1. OMG, O.M.: Omg unified modeling language (omg uml), superstructure, v2.1.2 (2007)
2. Mell, P., Scarfone, K., Romanosky, S.: Common vulnerability scoring system version 2.0, NIST and Carnegie Mellon University, 1st edn, June 2007
3. Mehedintu, A., Bulgiu, I., Pirvy, C.: Web-enabled Data Warehouse and Data Webhouse
4. Hernndez, P., Garrigs, J.: Model-driven development of multidimensional models from web log files, ER?10 Proceedings of the international conference on Advances in conceptual modeling: applications and challenges, pp. 170–179 (2010)
5. Liu, J., Hu Chaou, J., YuanHeJin: Application of Web Services on The Real-time Data Warehouse Technology (2010)
6. Kimball, R., Merz, R.: Le DATA WEBHOUSE:Analyser les comportements client sur le Web, Eyrolles Edition, 2000
7. Muralini, M., Kumar, T.V.S., Kanth, K.R: Simulating Secure Data Extraction in Extraction Transformation Loading (ETL) Processes, In Third UKSim European Symposium on Computer Modeling and Simulation, pp. 142–147 (2009)
8. Muralini, M., Kumar, T.V.S, Kanth, K.R.: Simulating: Secure ETL Process Model: An Assessment of Security in Different Phases of ETL, In Software Engineering Competence Center (2013)
9. Kiran, P., Sathish Kumar, S., Kavya, NP.: Modelling Extraction Transformation Load embedding Privacy Preservation using UML, Int. j. comput. Appl. (2012)
10. National Institute of Standards and Technology Special Publication 800–30, Risk Management Guide for Information Technology Systems, June 2001
11. National Institute of Standards and Technology Special Publication 800–53, Recommended Security Controls for Federal Information Systems, December 2007
12. National Institute of Standards and Technology Special Publication 800–55, Performance Measurement Guide for Information Security, July 2008
13. Cheng, P., Wang,L., Jajodia, S., Singhal, A.: Aggregating CVSS Base Scores for Semantics-Rich Network Security Metrics, In SRDS, 2012, pp. 31–40
14. Pengsu, C., Lingyu, W., Anoop, J.: Aggregating CVSS Base Scores for Semantics-Rich Network Security Metrics, pp. 31–40. IEEE, SRDS (2012)

15. Siv, H. Virginia, H., Franqueira, N.L., Erlend A. Engum.: Quantifying security risk level from CVSS estimates of frequency and impact, J. sys. softw. 83 (9), ISSN 0164-1212, pp. 1622–1634 (2010)
16. Mallek, H., Walha, A., Faiza, G.J., Gargouri, Faiez: ETL-Web process modeling, 8me edition de la confrence sur les Avancs des Systmes Dcisionnels, Hamamet Tunisia (2014)
17. Bellovin, S.: On the Brittleness of Software and the Infeasibility of Security Metrics, IEEE Security and Privacy (2006)
18. Thompson Lord Kelvin, W.: Electrical Units of Measurement,? Lecture at the Institution of Civil Engineers, London, 3 May 1883, Popular Lectures and Addresses, vol. 1, pp. 73–136 (1889)

Automated Negotiating Agent Based on Evolutionary Stable Strategies

Akiyuki Mori and Takayuki Ito

Abstract Bilateral multi-issue closed bargaining problems are critical in the research field of automated negotiations. In this paper, we propose a negotiating agent that is based on the expected utility value at the equilibrium point of an evolutionary stable strategy (ESS). Furthermore, we show the evaluation results of negotiating simulations, and demonstrate that our agent outperforms existing agents under the negotiation rules of the 2014 International Automated Negotiating Agents Competition (ANAC2014). Our paper has three contributions. First, our agent derives the expected utility of the ESS equilibrium points based on dividing a negotiation into two phases, and realizes the appropriate concession by the concession function that incorporates it. Second, it can reduce time discounts by a quick compromise based on an appropriate lower limit of concession value. Third, it can get beneficial results of negotiation simulations by the proposed concession function under various negotiation conditions.

1 Introduction

Automated negotiating agents are gaining a lot of interest in the field of multi-agent systems. Such agents negotiate instead of humans based on human preference information. They can build an appropriate consensus of large-scale bargaining problems that are difficult for humans to solve. There are much more applications that includes automated negotiating agent as e-commerce systems [1] and scheduling systems [2].

Concealing the negotiator's preference information is necessary from the viewpoint of privacy protection. Therefore, bilateral multi-issue closed bargaining problems (BMCBP) are critical in the research field of bargaining problems. The

A. Mori (✉) · T. Ito
Nagoya Institute of Technology, Gokisocho, Showa-ku Nagoya-shi, Aichi, Japan
e-mail: mori.akiyuki@itolab.nitech.ac.jp

T. Ito
e-mail: ito.takayuki@nitech.ac.jp

© Springer International Publishing Switzerland 2016
R. Lee (ed.), *Computer and Information Science 2015*,
Studies in Computational Intelligence 614, DOI 10.1007/978-3-319-23467-0_3

Automated Negotiating Agents Competition (ANAC), which has been held since 2010 as part of the International Joint Conference on Autonomous Agents and Multi-Agent Systems (AAMAS), is one BMCBP approach. The bargaining problems of ANAC have many constraints like a closed negotiations and time discounts that resemble real-life bargaining problems. In closed negotiations, negotiators are unable to know their opponent's utility.

Agents with a variety of negotiation strategies compete in ANAC. However, it is difficult for existing agents to continue to get beneficial negotiation results in various negotiation conditions because the effectiveness of existing negotiation strategies depends on those negotiation conditions. Negotiation strategy defines the approach of the negotiator's action to get a beneficial negotiation result. Beneficial negotiation results refer to reaching an agreement with high utility for a negotiator and high social welfare (sum of utility of negotiation participants).

We propose an automated negotiating agent that adapts to the variety of negotiation environments under the negotiation rules of ANAC2014 and explain about the overview of our negotiation strategy. First, our agent divides the negotiating flow into two phases: alternating offers phase (AOP) and final offer phase (FOP). Next, it defines the negotiation strategies in the FOP and estimates their acquisition utility of the strategies. After that, our agent analyzes the game matrix of the estimated acquisition utility as a strategy game and derives the expected utility value at the equilibrium point of the evolutionary stable strategy (ESS) [3]. The ESS used in the evolutionary game theory is an adaptive strategy and cannot be invaded by other strategies in the environments. Then our agent designs a concession function that contains the estimated expected utility of FOP in AOP.

Our agent's compromising method can provide beneficial decisions for the users of automated negotiation systems as a e-commerce systems that conceal the preference information of negotiators. The following are the main contributions of our paper:

- derives the expected utility of the ESS equilibrium points based on dividing a negotiation into two phases, and realizes the appropriate concession by the concession function that incorporates it.
- reduces time discounts by a quick compromise based on a appropriate lower limit of concession value.
- gets beneficial results of negotiation simulations by the proposed concession function under various negotiation conditions.

The remainder of this paper is organized as follows. Section 2 shows the negotiation environments of ANAC. Section 3 describes our agent's negotiation strategy, and Sect. 4 shows the results of experiments on negotiation simulations with ANAC2014 finalists. Section 5 discusses related works. Section 6 presents our conclusions and future works.

2 Negotiation Environment in ANAC

2.1 Overview of ANAC

ANAC is a BMCBP competition. Researchers from the worldwide negotiation community participate in ANAC. The following are the purposes of the competition:

- encourages the architecture of agents that can aptly negotiate under the various negotiation conditions.
- provides a benchmark for objectively evaluating different negotiation strategies of agents.
- researches learning methods, adaptive strategies, and models of opponents.
- collects front-line negotiating agents, and makes them accessible for the other research communities.

ANAC2014, the fifth ANAC, was held in AAMAS2014 in May 2014. 19 agents participated in ANAC2014 from all over the world and 10 agents had been selected from a qualifying tournament. The selected agents were ranked in a final tournament. In this paper, we compare by negotiation simulation our agent and the agents that participated in the final tournament of ANAC2014.

2.2 Negotiation Problems

2.2.1 Negotiation Domains

Since real-life bargaining problems are composed of multiple issues, negotiation domains are digitized with respect to each issue in ANAC. Agents can know the complete information of the negotiation domain.

2.2.2 Utility Function

A utility function, which defines the utility value of a bid (an agreement candidate), is based on constraints [4] and was used at ANAC2014. In BMCBP, a negotiator cannot obtain an opponent's utility function.

2.2.3 Negotiation Protocol

ANAC has a negotiation protocol that is based on alternating offers [5]. Alternating offers have been researched by many approaches, such as game theory [6, 7] and the heuristic settings of negotiations [8]. In alternating offers, negotiators can select one of the following three actions:

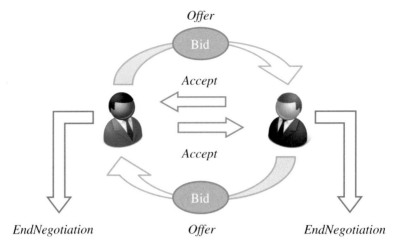

Fig. 1 Outlines of alternating offers

Offer: making a bid to an opponent. The first bidder is decided before the negotiation.
Accept: agreeing to a bid offered by the opponent. When a negotiator selects *Accept*,
each negotiators gets the utility value evaluated by its own utility function with a
time discounting. If the negotiator rejects its opponent's bid the negotiator makes
a counter offer for the opponent.
EndNegotiation: ending the negotiation without any agreement. For bilateral bargaining problems, a negotiation is closed if either negotiator selects *EndNegotiation*.
When it is selected, the negotiators get a reservation value that is discounted by
time.

For example, consider a negotiation between agents A and B. First, agent A offers
Bid_1 to agent B if agent A is the negotiator who first bids. After the offer from agent A,
agent B can select *Offer*, *Accept*, or *EndNegotiation*. Agent B selects *EndNegotiation*
when it wants to close the negotiation. Agent B selects *Accept* if it can accept Bid_1, and
the agents get utility values of Bid_1. When agent B rejects Bid_1, it offers Bid_2 to agent
A. After the offer from agent B, agent A can select *Offer*, *Accept*, or *EndNegotiation*
again. The negotiation continues like in this example. Figure 1 shows the relationships
between agents and three actions in alternating offers.

2.2.4 Limited Time

Since many real-life bargaining problems must be resolved in finite time, the negotiating time is limited to 180 s in ANAC. When the negotiating time exceeds this limit,
negotiators select *EndNegotiation*. In this paper, we normalize the elapsed time of
negotiation T, which is defined by formula (1). T_{max} means the maximum negotiation
time.

$$t = \frac{T}{T_{\text{max}}} \tag{1}$$

2.2.5 Discounted Utility

Acquisition utility values decrease with the elapsed time. The discounted utility by time is one constraint in bargaining problems that is discussed in existing researches [9]. In ANAC, the discounted utilities of bargaining problems are based on discount factor df ($0 < df \leq 1$). Each bargaining problem has its own discount factor. The utility function with discounted utility $U_D(\mathbf{s}, t)$ is defined by formula (2). $U(\mathbf{s})$ is a normalized utility function in [0,1], and $\mathbf{s}$ denotes the bid.

$$U_D(\mathbf{s}, t) = U(\mathbf{s}) \cdot df^t \tag{2}$$

2.2.6 Reservation Value

When either of the negotiators selects *EndNegotiation*, they get a fixed value called a reservation value. Each bargaining problem has its own reservation value. Moreover, reservation value RV is influenced by the discounted utility. A reservation value with discounted utility $RV_D(t)$ is defined by formula (3).

$$RV_D(t) = RV \cdot df^t \tag{3}$$

3 Negotiation Strategy Based on Estimated Expected Utility

3.1 Phases of Negotiation Flow

To determine the appropriate concession value in a negotiation, we focus on the limited time and divide the negotiating flow into two phases. Figure 2 shows the negotiating flow and the divided phases. Here we define an alternating offers phase (AOP) and a final offer phase (FOP). Our agent decides AOP's concession value by FOP's expected utility. In this section, we consider two phases and discuss their characteristics.

3.1.1 Alternating Offers Phase

Alternating offers phase (AOP) means the series of the flow from branch 1 to branch 4 in Fig. 2. If negotiators fail to get an opponent's agreement in AOP, the acquisition

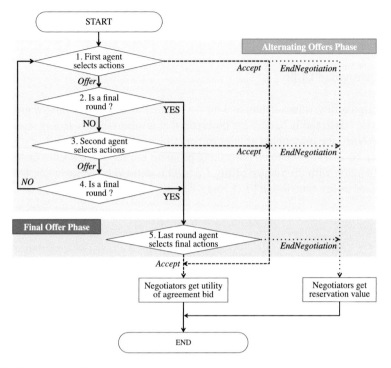

Fig. 2 Phases of negotiation flow

utility of the negotiators is decreased by a time discount. Discount factor per round δ is defined by formula (4). ε denotes the normalized interval time per round.

$$\delta = df^{\varepsilon} \qquad (4)$$

Under the negotiation rule of ANAC, a negotiator with the right to choose an action can select actions any time within the time limit. Therefore, ε is not fixed in ANAC. Since negotiators need to model their opponent's preferences within the time limit and offer as many bids as possible, we assume that ε is small sufficiently. When $\varepsilon \simeq 0$, δ is approximately 1.0. Consequently, the discounted utility per round is small due to δ, which is sufficiently small in AOP.

Our agent decides about the concession values based on the expected utility of FOP because it has a higher risk than AOP for the loss that occurs concerning agreement failures. However, the loss is accumulated and increases. When the discount factor is small, a rapid compromise is necessary to reduce the discounted utility.

3.1.2 Final Offer Phase

Final Offer Phase (FOP) is denoted by branch 5 in Fig. 2. An agreement failure means a negotiation failure in FOP. If a negotiation fails, the negotiators get utility based on the reservation values. The loss of negotiation failure X is defined by formula (5). C is the utility if a negotiator compromises, and $1.0 - \varepsilon$ denotes the compromise time to reach a final offer.

$$
\begin{aligned}
X &= df^{1.0-\varepsilon} \cdot C + df \cdot RV \\
&\simeq df \cdot (C - RV)
\end{aligned}
\tag{5}
$$

If C is sufficiently bigger than RV, X exhibits a great loss. For example, for $df = 1.0$, $RV = 0$, and $C = 0.5$. The loss of the utility value is 0.5. Negotiators need to compromise to prevent such losses. On the other hand, the important point is that an opponent has same contexts. If the opponent compromises to avoid negotiation failure, it can be exploited. However, if the opponent also considers the same way, the negotiation fails and the negotiators suffer a loss. The FOP situation resembles a game. In this example, its structure is not so much the ultimatum game as the chicken game. FOP's game structure is different from the ultimatum game where the negotiator cannot know the exact time of the final offer. However, unlike a general chicken game, the negotiation strategy's effectiveness depends on the negotiators' own preferences.

3.2 The Negotiation Game and Estimated Expected Utility in FOP

In this section, we define and analyze the negotiation game and estimate the expected utility value in FOP. In FOP, our agent classifies negotiation approaches into the following two strategies:

Hard-line Strategy (HS): not relaxing the agreement condition until the end of negotiation's end.

Compromise Strategy (CS): relaxing the agreement condition until just before the negotiation's end and building a consensus with a compromise proposal.

Table 1 shows the game matrix when the negotiators select HS or CS. When agents A and B select HS, the acquisition utility value of agent A is A_{11}, and the acquisition utility value of agent B is B_{11}. Agent A indicates our agent, and agent B indicates an opponent. Our agent derives an ESS based on Table 1. Moreover, our agent considers the expected utility of the ESS equilibrium point as FOP's expected utility. The estimated acquisition utilities of agent A: A_{11}, A_{12}, A_{21} and A_{22} are defined by formula (6a)–(6d).

Table 1 Negotiation game matrix in FOP

A	B	
	HS	CS
HS	(A_{11}, B_{11})	(A_{12}, B_{12})
CS	(A_{21}, B_{21})	(A_{22}, B_{22})

$$A_{11} = RV_A \cdot df_A^{1.0-\varepsilon} \tag{6a}$$

$$A_{12} = df_A^{1.0-\varepsilon} \tag{6b}$$

$$A_{21} = \begin{cases} C_A \cdot df_A^{1.0-\varepsilon} & (C_A \geq RV_A) \\ RV_A \cdot df_A^{1.0-\varepsilon} & (C_A < RV_A) \end{cases} \tag{6c}$$

$$A_{22} = p_c(A) \cdot A_{21} + (1.0 - p_c(A)) \cdot A_{12} \tag{6d}$$

df_A denotes the discount factor of agent A, RV_A means reservation value of agent A, and C_A means the acquisition utility value of agent A if it compromise. $p_c(A)$ means the compromise probability of agent A if agents A and B select CS. Moreover, our agent approximately regards ε as 0 and C_A as the utility value of the best bid offered to agent A. In formula (6a), if agents A and B select HS, the negotiation fails because the negotiators failed to compromise before the negotiation's end. Therefore, A_{11} is the reservation value with time discounts. In formulas (6b) and (6c), if CS is selected by either agents A or B, but not both, the agent that selects CS is exploited. Therefore, the utility value of the agent that selected CS is the estimated C with a time discount, and the utility value of the agent that selected HS is the estimated max utility 1.0 with a time discount. For formula (6d), CS does not relax the agreement condition until just before the final offer. If agents A and B select CS, the agent that quickly compromises is exploited. Therefore, the negotiators compete in the compromises time to exploit their opponent. If they attempt to compromise with each other after their opponent compromises, it is difficult to determine which agent compromised first. Consequently, $p_c(A)$ is 0.5 in this paper.

From the Table 1, the expected utility function of agent $AF_A(p, q)$ is defined as formula (7).

$$\begin{aligned} F_A(\mathbf{p}_A, \mathbf{p}_B) &= p_A \cdot p_B \cdot A_{11} + p_B \cdot (1 - p_B) \cdot A_{12} \\ &\quad + (1 - p_A) \cdot p_B \cdot A_{21} + (1 - p_A) \cdot (1 - p_B) \cdot A_{22} \\ &= p_A \cdot \{p_B \cdot (A_{11} - A_{21}) + (1 - p_B) \cdot (A_{12} - A_{22})\} \\ &\quad + p_B \cdot A_{21} + (1 - p_B) \cdot A_{22} \end{aligned} \tag{7}$$

$\mathbf{p}_X = (p_X, 1 - p_X)$ is the probability distribution of mixed strategy agent X, and p_X denotes the probability of selected HS by agent X. If $p_B^* \cdot (A_{11} - A_{21}) + (1 - p_B^*) \cdot (A_{12} - A_{22}) = 0$, the mixed strategy $\mathbf{p}_B^* = (p_B^*, 1 - p_B^*)$ is the ESS of agent B. Figure 3 shows the ESS equilibrium point. When agent B selects $\mathbf{p}_B^*$, $F_A(\mathbf{p}_A, \mathbf{p}_B^*)$

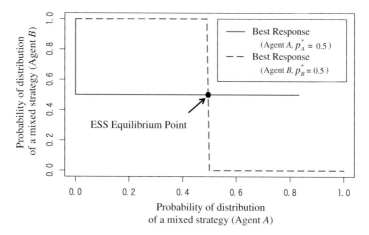

Fig. 3 ESS equilibrium point

is unaffected by $\mathbf{p}_A$. Thus, the expected utility of agent A at the equilibrium point of ESS $F_A(\mathbf{p}_A^*, \mathbf{p}_B^*)$ is defined by formula (8).

$$F_A(\mathbf{p}_A^*, \mathbf{p}_B^*) = p_B^* \cdot A_{21} + (1.0 - p_B^*) \cdot A_{22} \tag{8}$$

3.3 Design of Concession Function in AOP

The concession function defines the threshold value of the agreement. If the bid's utility value exceeds the concession function's value, negotiators accept the bid. The concession function of our agent has lower limit $L(t)$ which exceeds the expected utility of FOP with a time discount. The lower limit of the concession function of agent A $L_A(t)$ is defined by formula (9).

$$L_A(t) = F_A(\mathbf{p}_A^*, \mathbf{p}_B^*)/df_A^t \tag{9}$$

The concession function's value decreases from max utility value 1.0 to $L(t)$ because the negotiators want to decrease the concession value as much as possible. In this paper, the rate of the decrease in the concessions function depends on the time and is linear. When $df > 0$, a quick agreement is important to reduce the discounted utility. Therefore, our agent uses plural concession functions in accordance with the discount factor. Our agent's concession function $T(t)$ is designed as formula (10).

$$T(t) = \begin{cases} L(t) + (1.0 - L(t)) \cdot (1.0 - t) & (df = 1.0) \\ 1.0 - t/\alpha & (df < 1.0 \cap 1.0 - t/\alpha > L(t)) \\ L(t) & (df < 1.0 \cap 1.0 - t/\alpha \leq L(t)) \end{cases} \tag{10}$$

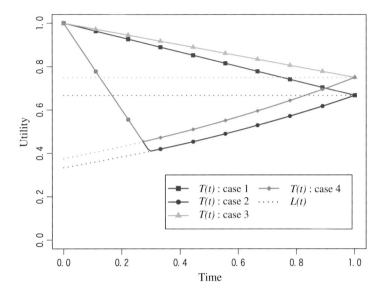

Fig. 4 $T(t)$ for cases 1–4

If $0.0 < df < 1.0$, $T(t)$ has minimum utility 0 when $t = \alpha$. The more a discount factor decreases, the more important a quick agreement is. Therefore, we decides $\alpha = df$. Moreover, $T(t) = L(t)$ when $T(t) < L(t)$. Figure 4 shows $T(t)$ in cases 1, 2, 3, and 4. The df and RV values of cases 1 to 4 are used in ANAC2014.

case 1: $df = 1.0$ $RV = 0.00$ $C = 0.5$
case 2: $df = 0.5$ $RV = 0.00$ $C = 0.5$
case 3: $df = 1.0$ $RV = 0.75$ $C = 0.5$
case 4: $df = 0.5$ $RV = 0.75$ $C = 0.5$

If $df > 0$ and $RV > 0$, as in case 4, $T(t)$ can be smaller than $RV_D(t)$. In this case, our agent selects *EndNegotiation* and decides the concession value according to $T(t)$ in AOP. However, agent A cannot obtain $\mathbf{p}_A^*$ because it is determined by B_{11}, B_{12}, B_{21}, and B_{22} as the opponent's private information. Thus, our agent selects CS to prioritize building a consensus in FOP.

4 Experimental Analysis

4.1 Negotiation Problems of ANAC2014

We evaluate our agent by running negotiations with 10 agents that are ANAC2014 finalists. In this simulation, we used the 12 bargaining problems from the ANAC-2014 for round-robin tournament. Table 2 shows the bargaining problems of ANAC2014. Each bargaining problem has two profiles that define a negotiator's utility function. $U(NBS)$ indicates the negotiator's utility value of the nash bargaining

Table 2 Negotiation problems of ANAC2014

ID	Issues	Profile	$U(NBS)$	df	RV
1	10	1	0.76	1.00	0.00
		2	0.92		
2	30	1	0.89		
		2	0.95		
3	50	1	0.98		
		2	0.90		
4	10	1	0.97	0.50	
		2	0.76		
5	30	1	0.95		
		2	0.95		
6	50	1	0.91		
		2	0.96		
7	10	1	0.85	1.00	0.75
		2	0.90		
8	30	1	0.90		
		2	0.92		
9	50	1	0.95		
		2	0.97		
10	10	1	0.90	0.50	
		2	0.88		
11	30	1	0.87		
		2	0.87		
12	50	1	0.95		
		2	0.87		

solution. We evaluated the agent's acquisition utility and social welfare and measure the ratio of acquisition utility value for the social welfare (exploiting ratio) and the discounted utility. Our experiment is same settings of ANAC2014. In this simulation, negotiators exchange own profile each other. Moreover, the number of trial of same negotiation setting is 10 times.

Table 3 shows the acquisition utility and the social welfare of the simulated agents. Both of our agent's acquisition utility and social welfare values are the highest in the simulated agents due to appropriate concessions. Therefore, it realized appropriate concessions in comparison with existing agents.

Table 4 shows the exploiting ratio and the discounted utility of the simulated agents. Both of our agent's acquisition utility and social welfare increase since its

Table 3 Acquisition and social welfare (negotiation problems of ANAC2014)

Agent name	Acquisition		Social welfare	
	Mean	SD	Mean	SD
Our agent	**0.833**	**0.162**	**1.610**	**0.226**
AgentM	0.766	0.143	1.601	0.255
Gangster	0.750	0.198	1.494	0.318
DoNA	0.743	0.152	1.460	0.284
GROUP2Agent	0.733	0.227	1.393	0.424
WhaleAgent	0.725	0.212	1.462	0.374
E2Agent	0.722	0.216	1.437	0.375
kGA_gent	0.700	0.303	1.330	0.553
AgentYK	0.691	0.203	1.392	0.394
BraveCat v0.3	0.681	0.173	1.409	0.354
ANAC2014Agent	0.622	0.203	1.392	0.397

Table 4 Exploiting ratio and discounted utility (negotiation Problems of ANAC2014)

Agent name	Exploiting ratio		Discounted score	
	Mean	SD	Mean	SD
Our agent	**0.517**	**0.079**	**0.025**	**0.155**
AgentM	0.479	0.049	0.031	0.174
Gangster	0.499	0.064	0.066	0.248
DoNA	0.515	0.086	0.014	0.118
GROUP2Agent	0.529	0.066	0.105	0.306
WhaleAgent	0.492	0.060	0.077	0.266
E2Agent	0.500	0.068	0.093	0.290
kGA_gent	0.521	0.043	0.081	0.273
AgentYK	0.498	0.052	0.102	0.302
BraveCat v0.3	0.487	0.059	0.083	0.275
ANAC2014Agent	0.450	0.076	0.049	0.216

exploiting ratio is high, and its discounted utility is small in the agents from the tournament. However, a high exploiting ratio doesn't necessarily mean good negotiation results. For example, GROUP2Agent and kGA_gent have higher exploiting ratios than our agent. Nevertheless, our agent's score is better than GROUP2Agent and kGA_gent, since their discounted utility increased because their agreement conditions are too strict. Therefore, it is important for negotiators to decrease their discounted utility and to exploit the opponent to get good negotiation results.

4.2 Various Negotiation Problems

ANAC2014 supposes that bargaining problems have $df = 1.0$ or 0.5 and $RV = 0.0$ or 0.75. In this paper, we use various bargaining problems to evaluate the adaptivity of our agent. We use a 10 issues negotiation problem from ANAC2014 (referred to as 10-issues). We created the 16 bargaining problems, which are 10-issues with $df = 0.25, 0.5, 0.75$ or 1.0 and $RV = 0.0, 0.25, 0.5$, or 0.75. df and RV are the same values between the negotiators in these bargaining problems; we evaluated the agent's acquisition utility and social welfare as in our previous experiment.

Table 5 shows the acquisition utility and the social welfare of the simulated agents. Both of our agent's acquisition utility and social welfare values are the highest like in the previous experiment.

Table 6 shows the exploiting ratio and the discounted utility of the simulated agents. Our agent's exploiting ratio is less than 0.5 because our agent quickly compromised. Even if an agent accepts an adverse condition bid, the acquisition utility and social welfare increase due to a small discounted utility. Our agent's compromise is appropriate.

Comparing our two experiments, the ranking of the agent scores has changed because the negotiation conditions changed. Thus, agents depending on negotiation conditions cannot adapt to the fluctuations of negotiation environments. The effectiveness of existing agent negotiation strategies also depends on the negotiation conditions. On the other hand, since our agent's rank was constantly high in both experiments, the effectiveness of our agent's negotiation strategy is not dependent on the negotiation conditions. Our agent is more adaptable to the fluctuations of negotiation conditions than existing agents.

Table 5 Acquisition and social welfare (various negotiation problems)

Agent name	Acquisition		Social welfare	
	Mean	SD	Mean	SD
Our agent	**0.742**	**0.155**	**1.498**	**0.177**
DoNA	0.740	0.229	1.307	0.346
Gangster	0.671	0.201	1.347	0.253
GROUP2Agent	0.666	0.209	1.350	0.366
BraveCat v0.3	0.646	0.220	1.265	0.387
AgentYK	0.637	0.235	1.210	0.382
AgentM	0.633	0.207	1.322	0.372
E2Agent	0.630	0.252	1.188	0.410
WhaleAgent	0.622	0.256	1.221	0.408
kGA_gent	0.580	0.286	1.088	0.482
ANAC2014Agent	0.535	0.151	1.415	0.210

Table 6 Exploiting ratio and discounted utility (various negotiation problems)

Agent name	Exploiting Ratio		Discounted Utility	
	Mean	SD	Mean	SD
Our agent	**0.494**	**0.088**	**0.039**	**0.194**
DoNA	0.568	0.117	0.026	0.160
Gangster	0.493	0.093	0.095	0.294
GROUP2Agent	0.500	0.101	0.083	0.275
BraveCat v0.3	0.512	0.078	0.136	0.343
AgentYK	0.525	0.085	0.179	0.384
AgentM	0.482	0.091	0.111	0.315
E2Agent	0.523	0.079	0.157	0.364
WhaleAgent	0.500	0.090	0.129	0.335
kGA_gent	0.521	0.068	0.066	0.249
ANAC2014Agent	0.379	0.094	0.026	0.158

5 Related Work

Many automated negotiating agents have been proposed in ANAC. AgentK [10], which was the best agent in ANAC2010. Its negotiation strategy controls the compromise values based on its opponent's bidding history. AgentK estimates the utility space and the negotiating posture of opponents from their bidding history. Hard-Headed [11] was the best agent in ANAC2011 [12]. It adjusts the parameters of the concession function according to a discount factor and time. The Fawkes [13] was the best agent in ANAC2013. It estimates optimal concession values by discrete wavelet prediction [14]. AgentM was the best agent in ANAC2014. Its negotiation strategy is based on the time and the utility differences of the offered bids. The utility difference denotes the difference between the highest and smallest utilities.

The problem of the existing agents is their effectiveness under various negotiation conditions. In tournaments of existing agents, ranking by agents' acquisition utility was changed by the negotiation conditions. Also, many existing agents decide concession values based on evaluations of the bidding history of opponents by their own utility functions. However, negotiators cannot obtain a correlation of mutual preferences. Thus, existing negotiation strategies, which are based on evaluating the bidding history of opponents by their own utility functions, are not appropriate.

Our agent uses a concession function that incorporates the estimated expected utility of negotiation. Compared to the existing agents, the effectiveness of our agent's strategy is not dependent on the correlation of mutual preferences and opponent's negotiation strategies. Our agent can compromise appropriately and get a beneficial negotiation results under various negotiation conditions.

6 Conclusion

We proposed an automated negotiating agent to estimate the appropriate concession values. The concession function of our agent is designed based on the expected utility value at the ESS equilibrium point. In the evaluation experiments, we simulated negotiations with our agent and the ANAC2014 agents. We showed that our agent's acquisition utility and social welfare were the highest among the simulated agents. When the negotiating conditions were changed, our agent's acquisition utility and social welfare remained the highest among the simulated agents. Thus, our agent's concession function is appropriate.

Future work will experiment with more various negotiation conditions to prove the efficiency of our strategy. For the speed of compromises, we need to verify the effectiveness of non-linear functions.

References

1. Kraus, S.: Strategic Negotiation in Multiagent Environments. MIT Press, Cambridge (2001)
2. Sen, S., Durfee, E.H.: On the design of an adaptive meeting scheduler. In: Proceedings of the Tenth IEEE Conference on AI Applications, pp. 40–46. IEEE (1994)
3. Smith, J.M.: Evolution and the Theory of Games. Cambridge University Press, Cambridge (1982)
4. Ito, T., Hattori, H., Klein, M.: Multi-issue negotiation protocol for agents: exploring nonlinear utility spaces. IJCAI **7**, 1347–1352 (2007)
5. Rubinstein, A.: Perfect equilibrium in a bargaining model. Econom.: J. Econom. Soc. 97–109 (1982)
6. Shaked, A., Sutton, J.: Involuntary unemployment as a perfect equilibrium in a bargaining model. Econom.: J. Econom. Soc. 1351–1364 (1984)
7. Osborne, M.J., Rubinstein, A.: Bargaining and markets (economic theory, econometrics, and mathematical economics) (1990)
8. Faratin, P., Sierra, C., Jennings, N.R.: Negotiation decision functions for autonomous agents. Robot. Auton. Syst. **24**(3), 159–182 (1998)
9. Fatima, S., Wooldridge, M., Jennings, N.R.: An analysis of feasible solutions for multi-issue negotiation involving nonlinear utility functions. In: Proceedings of the 8th International Conference on Autonomous Agents and Multiagent Systems. International Foundation for Autonomous Agents and Multiagent Systems, vol. 2, pp. 1041–1048 (2009)
10. Kawaguchi, S., Fujita, K., Ito, T.: Agentk: compromising strategy based on estimated maximum utility for automated negotiating agents. New Trends in Agent-Based Complex Automated Negotiations, pp. 137–144. Springer, New York (2012)
11. van Krimpen, T., Looije, D., Hajizadeh, S.: Hardheaded. Complex Automated Negotiations: Theories, Models, and Software Competitions, pp. 223–227. Springer, New York (2013)
12. Baarslag, T., Fujita, K., Gerding, E.H., Hindriks, K., Ito, T., Jennings, N.R., Jonker, C., Kraus, S., Lin, R., Robu, V., et al.: Evaluating practical negotiating agents: results and analysis of the 2011 international competition. Artif. Intell. **198**, 73–103 (2013)
13. Baarslag, T.: What to bid and when to stop. Ph.D. dissertation, TU Delft, Delft University of Technology (2014)
14. Chen, S., Weiss, G.: An efficient and adaptive approach to negotiation in complex environments. In: ECAI, pp. 228–233 (2012)

Architecture for Intelligent Transportation System Based in a General Traffic Ontology

Susel Fernandez, Takayuki Ito and Rafik Hadfi

Abstract Intelligent transportation systems are a set of technological solutions used to improve the performance and safety of road transportation. A crucial element for the success of these systems is that vehicles can exchange information not only among themselves but with other elements in the road infrastructure through different applications. For the success of this exchange of information, a common framework of knowledge that allows interoperability is needed. In this paper an ontology-based system to provide roadside assistance is proposed, providing drivers making decisions in different situations, taking into account information on different traffic-related elements such as routes, traffic signs, traffic regulations and weather elements.

Keywords Intelligent transportation systems · Ontology · Reasoning · Agents

1 Introduction and Related Work

Today, it is known as Intelligent Transportation Systems, the set of applications and technological systems created with the aim of improving safety and efficiency in road transport. These systems allow to control, manage and monitoring the different elements of roads.

The continuing evolution of intelligent transportation systems has ushered in a new era of interconnected intelligent systems, which certainly has been a quantitative leap in safety of road transport. These systems enable the exchange of information between different applications, and the subsequent analysis to improving the safety of drivers and eases travel and comfort in road travel.

S. Fernandez (✉)
University of Alcala. Alcalá de Henares, Madrid, Spain
e-mail: susel.fernandez@uah.es

T. Ito · R. Hadfi
Nagoya Institute of Technology, Nagoya, Japan
e-mail: ito.takayuki@nitech.ac.jp

R. Hadfi
e-mail: rafik@itolab.nitech.ac.jp

© Springer International Publishing Switzerland 2016
R. Lee (ed.), *Computer and Information Science 2015*,
Studies in Computational Intelligence 614, DOI 10.1007/978-3-319-23467-0_4

Due to its high degree of expressiveness, the use of ontologies is crucial to ensure greater interoperability among software agents and different applications involved in intelligent transportation systems. Ontologies provide a common vocabulary in a given domain and allow defining, with different levels of formality, the meaning of terms and the relationships between them [1]. Ontologies facilitates the design of exhaustive and rigorous conceptual schemas to allow communication and information exchange between different systems and institutions.

There are some previous works focused on ontology for road transportation systems. In [2] an ontology to represent traffic in highways has been developed. Its aim was the construction of reliable Traffic Information System providing information about roads, traffic, and scenarios related to vehicles in the roads. It also provides ways to the Traffic Information System to analyze how critical a specific situation is. For example, an ambulance may need to know about the congestion status of a toll plaza. Requesting this information is critical if the ambulance is moving to the scene of an accident. On the other hand, if a common vehicle is moving through a road without hurry, then its information requested is not critical.

In [3] they proposed a high-level representation of an automated vehicle, other vehicles and their environment, which can assist drivers in taking "illegal" but practical relaxation decisions (for example when a car damaged does not allow the circulation, take the decision to move to another lane crossing a solid line and overtake the stopped car, if the other lane is clear). This high-level representation includes topological knowledge and inference rules, in order to compute the next high-level motion an automated vehicle should take, as assistance to a driver.

In [4] an ontology-based spatial context model was proposed. The work takes a combined approach to modeling context information utilized by pervasive transportation services: the Primary-Context Model facilitates interoperation across independent Intelligent Transportation Systems, whereas the Primary-Context Ontology enables pervasive transportation services to reason about shared context information and to react accordingly. The independently defined, distributed information is correlated based on its primary-context: location, time, identity, and quality of service. The Primary-Context Model and Ontology have been evaluated by modelling a car park system for a smart parking space locator service.

The work proposed in [5] is an approach to create a generic situation description for advanced driver assistance systems using logic reasoning on a traffic situation knowledge base. It contains multiple objects of different type such as vehicles and infrastructure elements like roads, lanes, intersections, traffic signs, traffic lights and relations among them. Logic inference is performed to check and extend the situation description and interpret the situation e.g. by reasoning about traffic rules. The capabilities of this ontological situation description approach are shown at the example of complex intersections with several roads, lanes, vehicles and different combinations of traffic signs and traffic lights.

In the work in [6] an ontology for traffic management is proposed, adding certain concepts of traffic to general sensor ontology A3ME [7]. The added concepts are specializations of position, distance and acceleration sensor classes, and the different actions that take place in the car motions. The ontology is developed in OWL, using the JESS reasoner with SWLR [8] rules.

In [9] an ontology-based Knowledge Base, which contains maps and traffic regulations, was introduced. They can aware over speed situations and make decisions at intersections in comply with traffic regulations, but they did not consider important elements such as traffic signals and weather conditions.

Most of the works found in the literature focus on describing very specific traffic situations such as finding parking, actions of emergency vehicles and intersection situations. But none of them is general and expressive enough to encompass a wide variety of traffic situations. Therefore it's necessary to develop ontologies in the domain of road traffic expressive enough to describe any traffic situation. The ontologies should be richer with respect to the various sensor inputs, and use it to drastically improve the general routing mechanism.

This paper presents an ontology-based system for road transportation management, with the aim of providing driver assistance in different traffic situations. The architecture is based on a centralized mechanism for smart routing. The developed ontology manages the knowledge related to vehicles and environmental elements that can influence road traffic such as infrastructure elements, weather conditions and traffic regulations in Japan.

The paper is organized as follows. Section 2 presents the architecture of the ontology–based system. In Sect. 3 the case of study with different traffic scenarios are explained. Finally the conclusions and lines of future work are summarized in Sect. 4.

2 Ontology-Based System Architecture

The intelligent system for road traffic assistance consists of three layers, as shown in Fig. 1.

At the bottom is the ontological layer, where we have developed the ontology of the domain of road traffic in OWL, the second is reasoning layer, in which we have used the reasoner Pellet, and finally in the upper layer are the agents that access to information of the ontology through SPARQL queries.

2.1 Ontology Layer

In the ontology layer of the system, an ontology that relates the different road traffic entities has been developed. The ontology was implemented in OWL-RDF language 0 using the protégé tool 0.

Fig. 1 System architecture

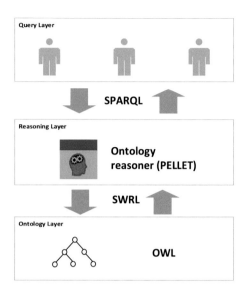

For better understanding, we present the knowledge in the traffic ontology divided in two groups of interrelated concepts. The first group contains the elements related to the vehicles, and the second group the elements related to road infrastructure. We explain in detail the two groups below.

The main group is related to vehicles. The concepts of this group are shown in Fig. 2. The figure shows the taxonomy of vehicles, which can be classified into: commercial vehicles, public vehicles (bus and taxi), private vehicles (car, bicycle and motorbike) and priority vehicles (ambulances, fire trucks and police cars). Different relationships between vehicles and other entities are defined also in this group. Some of these entities are: location, showing the exact location (latitude and longitude) of a vehicle, route point or infrastructure item; information about drivers and the vehicle's types which they can drives by license.

One of the most important issues in this group is that each vehicle has associated a set of actions, which may vary depending on the route and traffic signals found, and a set of warnings depending on the weather situation in the area.

Regarding sensors, these can be located not only on vehicles but also on different parts of the infrastructure, such as bridges, roads, signs, etc. Various types of sensors have been defined in the ontology such as: vibration, acceleration, humidity, temperature, etc.

Figure 3 shows the second group, which organizes the elements related to road infrastructure. In this group the most important concept represents the roads, which in Japan are classified as local roads, prefectural roads, national highways and national expressways.

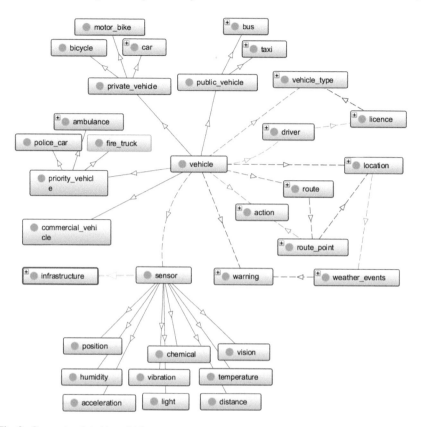

Fig. 2 Concepts related to vehicles

For better management of traffic situations we divided the roads into segments, connected through intersections. Each segment contains lanes, and different signs such as stop signs or speed control, traffic lights or road markings are in each lane. Each signal has an action associated following the Japanese traffic regulations.

2.2 Reasoning Layer

A crucial aspect when working with ontologies is the mechanism of reasoning, which in Artificial intelligence is simply the ability to obtain new knowledge from knowledge already available using inference strategies. To reason with ontologies, mainly three techniques are used: reasoning with First Order Logic, reasoning with Description Logic and reasoning with Rules.

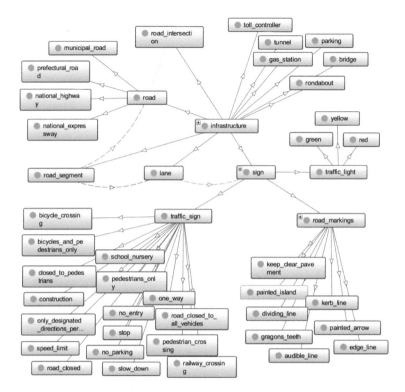

Fig. 3 Concepts related to infrastructure

In this work we used the reasoner Pellet 0, which is a tool for reasoning with ontologies, which supports the three types of reasoning. PELLET is implemented in Java; it is freely available and allows checking the consistency of the ontology.

The rules of reasoning in the traffic ontology have been developed using the Semantic Web Rule Language (SWRL) 0. In this ontology, SWRL rules are used to define different traffic regulations and the different actions that a driver can take, according to the current situation of the road. Here are some simple examples of rules defined in the traffic ontology.

The following are two simple rules that allow the reasoner inferring transitivity regarding to the location of the traffic elements. This means that if a traffic element (e.g. a vehicle or traffic signal) is located in a lane, and that lane is located in a road segment, then the traffic element of is also located in that road.

```
isOnLane(?c,?l),
isOnSegment(?l,?s)->
isOnSegment(?c,?s)

isOnSegment(?c,?s),
isOnRoad(?s,?r)-> isOnRoad(?c,?r)
```

The following example is the traffic rule applied when the priority vehicle is behind a private vehicle in the same lane. In this case, the action that the private vehicle must take is give way to the priority vehicle.

```
priority_vehicle(?prv), private_vehicle(?pv),
isOnLane(?prv,?l), isOnLane(?pv,?l),
is_behind(?prv,?pv)

-> has_action_value(?pv,yield_action)
```

The following is one of the most important rules of the traffic ontology. This rule aims to locate a traffic item on its corresponding road segment, considering its location (latitude and longitude) and the location of the start point and end point of each segment.

```
road_segment(?s), has_begining_location(?s,?b_loc),
has_ending_location(?s,?e_loc), has_location(?c,?loc),
has_latitude(?b_loc,?b_loc_lat),
has_latitude(?e_loc,?e_loc_lat),
has_latitude(?loc,?lat),
has_longitude(?b_loc,?b_loc_long),
has_longitude(?e_loc,?e_loc_long),
has_longitude(?loc,?long),
greaterThanOrEqual(?lat,?b_loc_lat),
greaterThanOrEqual(?long,?b_loc_long),
lessThanOrEqual(?lat,?e_loc_lat),
lessThanOrEqual(?long,?e_loc_long)-> isOnSegment(?c,?s)
```

2.3 Query Layer

The top layer of the system is the query layer. Here the different agents perform their tasks using the information stored in the ontology. As ontology query language has been used SPARQL [13].

Here is a simple example of a SPARQL query that returns the list of vehicles that are located in a given road.

```
PREFIX  rdfs:  <http://www.w3.org/2000/01/rdf-schema#>

PREFIX  owl:   <http://www.w3.org/2002/07/owl#>

PREFIX  xsd:   <http://www.w3.org/2001/XMLSchema#>

PREFIX  rdf:   <http://www.w3.org/1999/02/22-rdf-syntax-
ns#>

PREFIX  Traffic_ontology:
<http://www.semanticweb.org/ontologies/Traffic_ontology#>

SELECT  ?Vehicle ?Color ?Model ?Manufacturer ?Registra-
tion_Number ?Type ?Lane

WHERE

  { ?Vehicle rdf:type Traffic_ontology:vehicle .

    ?Vehicle Traffic_ontology:isOnRoad Traf-
fic_ontology:municipal_road1 .

    ?Vehicle Traffic_ontology:has_color ?Color .

    ?Vehicle Traffic_ontology:has_model ?Model .

    ?Vehicle Traffic_ontology:has_manufacturer ?Manufac-
turer .

    ?Vehicle Traffic_ontology:has_registration ?Registra-
tion_Number .

    ?Vehicle Traffic_ontology:has_vehicle_type ?Type .

    ?Vehicle Traffic_ontology:isOnLane ?Lane

  }
```

The following query returns all points associated with a route of a vehicle and the action that should be done to go from one point to the next, considering its location on the map. This query is very simple, taking into account that in the ontology each route point is related to the next one through an action (turn left, turn right or go straight), and every action depends on the type of relationship (*isAtNorthOf*, *isAtSouthOf*, *isAtWestOf*, *isAtEastOf*) that connects the segments in which the points of the route are.

There is a SWRL rule in ontology that associates an action or movement to move from one point to the next point on the route according to the relationship between the segments in which each route point is located.

```
PREFIX   rdfs:  <http://www.w3.org/2000/01/rdf-schema#>

PREFIX   owl:   <http://www.w3.org/2002/07/owl#>

PREFIX   xsd:   <http://www.w3.org/2001/XMLSchema#>

PREFIX   rdf:   <http://www.w3.org/1999/02/22-rdf-syntax-
ns#>

PREFIX   Traffic_ontology:
<http://www.semanticweb.org/ontologies/Traffic_ontology#>

SELECT   ?Point ?Action_to_next_point ?Next_point

WHERE

  { Traffic_ontology:car Traffic_ontology:has_route ?Route
  .

    ?Route Traffic_ontology:has_point ?Point .

    ?Point Traffic_ontology:has_desired_action ?Ac-
tion_to_next_point .

    ?Point Traffic_ontology:has_next_point ?Next_point

  }

ORDER BY ?Point
```

3 Traffic Situation Example

We have tested the expressiveness of ontology, querying on specific traffic situations. In this section we present an example of a simple traffic situation consisting of four roads and four intersections. As seen in Fig. 4 each road is divided into three segments, with two lanes each.

We have two vehicles (car1 and car2), of which its location, speed and path they want to follow are known. Each route has a set of points and each route point has a location (latitude and longitude) and the information about the next point in the route. The figure shows that it is raining in one of the intersections represented. We want to know the next action that the driver should take, given the vehicle's position, the chosen route and the traffic signs located along the route. We also want to know if there is any advice regarding the weather situation along the route.

In this work we have called *Desired_action* to the action that the driver wishes to take to move from one point to the next point along the route, regardless of traffic signals; *Next_Action* is the action that the driver should really take in each point considering only traffic signals. We have defined warnings for different weather situations that can be found on the route, such as rain, snow, fog, wind, etc. Each of these warnings is associated with a set of advices to facilitate the motion in these conditions.

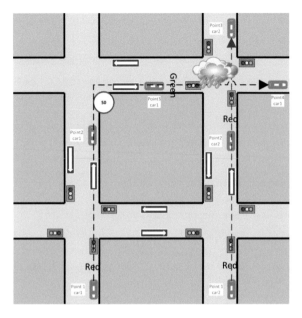

Fig. 4 Traffic example using routes

Table 1 Actions of the car1 through the route

Route 1	Next_Action	Desired_action	Advice
Point 1	Stop	Go_Straight	–
Point 2	Slow_down	Turn_ Right	–
Point 3	Go_Straight	Go_Straight	Turn on the lights
			Avoid puddles or flowing water
Point 4	–	–	–

In Table 1, the results obtained by the system for car1 at the different points of the chosen route (route 1) are shown. Route 1 is composed of four points which can be seen in Fig. 4, through a discontinuous black arrow. The speed of the car1 is 60 km/h.

Starting from the position of the vehicle at each point, the location of traffic signals and route, the actions are inferred by reasoning applying different rules of the ontology, in each of the steps outlined below:

1. Locate on which segment of the road are the vehicle and the next point on the route. This is done by considering the position (latitude and longitude) and the coordinates of the start and end points of each segment.
2. Choosing the desired action to go from one point to the next depending on the type of connection between the segments in which the points are located. For example, if the vehicle is in segment1 and the next point on the route is in the

segment2; and segment2 is located at east of segment1, then the action that the vehicle should take to go from point 1 to point 2 is to turn right. This rule would be expressed as follows:

```
isOnSegment(?vehicle,?s1), isOnSegment(?next_route_point,?s2),
isAtEastOf(?s2,?s1)->Desired_action(?vehicle, ?Turn_right)
```

3. Choosing the next action to be taken by the vehicle, taking into account only the traffic signs. This is the same action associated with the following traffic signal located in the segment where the vehicle is located. For example, if the vehicle is in a segment with a stop sign, the action that the driver should take is to stopping. This rule would be expressed as follows:

```
isOnSegment(?vehicle,?s1),
isOnSegment(?sign,?s1),has_action_value(?sign,?action)-
>Next_Action(?vehicle,?action)
```

4. If there is any special weather condition in the next segment of the path, then it is assigned to the vehicle the warning corresponding with that weather condition.

As seen in the table, the initial position of the vehicle is point 1, then the desired action to go from point 1 to point 2 through the route is *Go_Straight*, however the next action that the driver must take is stopping because there is a red light in that segment. To go from point 2 to point 3 the same procedure is performed, therefore the desired action is *Turn_Right*. However respect to the next action action, once detecting that a speed control signal in the segment, the vehicle speed is compared with the speed limit and as is greater (the speed limit is 50 km/h and the vehicle speed is 60 km/h), then the next action should be *Slow_down*. To go from point 3 to point 4 the desired action is *Go_straight*, and as there is a green light in that segment, then the next action is *Go_Straight* too. It's raining at the intersection connecting these two segments of the route and therefore the advices for the rain warning are selected in this point. These advices are: *Turn on the lights* and *Avoid puddles or flowing water*. Point 4 is the end of the route and therefore there is no action at this point.

Table 2 shows the actions related to the route of the car2. As shown in the Fig. 4, route 2 (shown by the discontinuous red arrow) is simpler than the other. By following the above steps, to go from point 1 to point 2, the vehicle 2 should go straight, but

Table 2 Actions of the car2 through the route

Rute 2	Next_Action	Desired_action	Advice
Point 1	Stop	Go_Straight	–
Point 2	Stop	Go_Straight	Turn on the lights
			Avoid puddles or flowing water
Point 3	–	–	–

is a red light, so it must stop. To go from point 2 to point 3, the same occurs, but because of it's raining at the intersection connecting the two segments of the route, the advices for the rain warning are selected. Finally there is no action in point 3 because it is the end of the route.

Overall, the results show that the ontology is quite expressive in terms of traffic signals, routes and traffic rules in Japan. However it is not expressive enough to take into account the behavior of drivers. The ontology allows inferring knowledge related to the weather from sensor data, but there are infrastructure sensors that measure other important data such as the crowd flow and traffic flow, which have not yet been taken into account in the ontology. The processing of data from those sensors will improve in the route optimization.

4 Conclusions

In this paper an ontology-based system for road transportation management is presented. The principal aim of this work is providing driver assistance in different traffic situations, taking into account the route, the weather and the traffic regulations in Japan.

The architecture of the proposed system consists of three interconnected layers. In the lower layer, an ontology that includes the most important concepts in the domain of road traffic and their relationships was designed. In the second layer is the reasoner, which performs the function of inferring new knowledge using the different rules and axioms of the ontology. Finally in the upper layer, agent tasks are performed through the different queries to the ontology depending on the specific traffic situations.

The expressiveness of ontology has been tested through queries in different traffic situations involving various signals and the traffic rules in Japan. The results of the tested scenarios have been satisfactory, but it is still need to enrich the ontology linking more knowledge. Therefore as future work we intend to continue to improving the expressiveness of the ontology, with processing data from more sensors located in the infrastructure, for example on bridges, roads, rivers, tunnels. Those sensors could measure crowd flow, traffic flow and many other parameters which are important in traffic optimization. We also intend to improve the expressiveness of the ontology adding information concerning the behavior of drivers, due to its importance throughout the road driving process. Finally we plan adding SWRL rules that describe multiples negotiation mechanisms between agents in different traffic scenarios.

Acknowledgments This work was partially supported by Research and Development on Utilization and Fundamental Technologies for Social Big Data by NICT (National Institute of Information and Communications Technology), and the Fund for Strengthening and Facilitating the National University Reformations by Ministry of Education, Culture, Sports, Science, and Technology, Japan.

References

1. Studer, R., Benjamins, R., Fensel, D.: Knowledge engineering: principles and methods. Data Knowl. Eng. **25**(1–2), 161–197 (1998)
2. Sérgio G., Ícaro S.: An ontology for a fault tolerant traffic information system. In: 22nd International Congress of Mechanical Engineering (COBEM 2013). 2013 November 3–7 Ribeirão Preto, SP, Brazil
3. Evangeline, P., Philippe M., Fawzi Nashashibi.: An ontology-based model to determine the automation level of an automated vehicle for co-driving. FUSION 2013, pp. 596–603
4. Lee, D, Meier, R.: Primary-context model and ontology: a combined approach for pervasive transportation services. In: First International Workshop on Pervasive Transportation Systems (PerTrans 2007), with the Fifth Annual IEEE International Conference on Pervasive Computing and Communications Workshops, PerCom 07, pp. 419–424
5. Michael Hülsen, J. Marius, Z., Christian, W.: Traffic Intersection situation description ontology for advanced driver assistance. In: 2011 IEEE Intelligent Vehicles Symposium (IV) Baden-Baden, Germany 5–9 June 2011
6. A.J. Bermejo, J. Villadangos, Astrain, J.J., Cordoba, A.: Ontology based road traffic management. In: Fortino, G., et al. (eds.) Intelligent Distributed Computing VI, SCI 446, pp. 103–108
7. Herzog, A., Jacobi, D., Buchmann, A.: A3ME-An agent-based middleware approach for mixed mode environments. In: Proceeding of Second International Conference on Mobile Ubiquitous Computing, Systems, Services and Technologies (UBICOMM 2008), Valencia, Spain, 29 September-4 October 2008; pp. 191–196
8. Horrocks, I., Patel-Schneider, P.F., Boley, H., Tabet, S., Grosof, B., Dean, M.: SWRL: a semantic web rule language combining OWL and RuleML, submission to W3C (2004). http://www.w3.org/Submission/SWRL/
9. Zhao, L., Ichise, R., Mita, S., Sasaki, Y.: Ontologies for Advanced Driver Assistance Systems
10. Dean, M., Schreiber, G.: OWL Web Ontology Language Reference. (2004). http://www.w3.org/TR/2004/REC-owl-ref-20040210/
11. Protégé: http://protege.stanford.edu/
12. Pellet: http://clarkparsia.com/pellet/
13. SPARQL: http://sparql.org/

Optimization of Cross-Lingual LSI Training Data

John Pozniak and Roger Bradford

Abstract The technique of latent semantic indexing (LSI) is widely employed in applications to provide information retrieval, categorization, clustering, and discovery capabilities. In these applications, the key relevant feature of the technique is the ability to compare objects (such as documents and queries) based on the semantics of their constituents. These comparisons are carried out in a high-dimensional vector space. That space is generated based on an analysis of occurrences of features in items of a training set. In the LSI literature there are multiple references to the fact that training items should be selected that are similar in content to the items to be dealt with in the application. This paper presents a principled approach for making such selection. We present test results for the technique for cross-lingual document similarity comparison. The results demonstrate that, at least for this use case, employment of the technique can have a dramatic beneficial effect on LSI performance.

Keywords Latent semantic indexing · LSI · LSI training · LSI optimization · Training set optimization · Cross-lingual LSI

1 Introduction

Latent semantic indexing is a versatile dimensionality reduction and data analysis technique that is employed in a wide variety of categorization, clustering, search, and information discovery applications. The LSI algorithm generates a high-dimensional vector space based on analysis of a corpus of training data supplied for a given problem. Numerous studies have shown that the size and composition of that training corpus can have a dramatic effect on the performance of LSI for that problem.

J. Pozniak
Leidos, Chantilly, VA 20151, USA
e-mail: john.e.pozniak@leidos.com

R. Bradford (✉)
Maxim Analytics, Reston, VA 20190, USA
e-mail: rogerbbradford@gmail.com

© Springer International Publishing Switzerland 2016
R. Lee (ed.), *Computer and Information Science 2015*,
Studies in Computational Intelligence 614, DOI 10.1007/978-3-319-23467-0_5

Although many interesting results have been reported, no general framework for creating effective LSI training spaces has emerged. In this paper we present a simple approach for selecting effective training data. In the following sections we present a brief description of LSI, a review of previous work, and motivation for the proposed technique. Experimental results for an English-Farsi cross-lingual document comparison problem demonstrate the effectiveness of the approach.

2 LSI Technique

The LSI technique applied to a collection of documents consists of the following primary steps [1]:

1. A matrix A is formed, wherein each row corresponds to a term that appears in the documents, and each column corresponds to a document. Each element $a_{m,n}$ in the matrix corresponds to the number of times that the term m occurs in document n.
2. Local and global term weighting is applied to the entries in the term-document matrix.
3. Singular value decomposition (SVD) is used to reduce this matrix to a product of three matrices:

$$A = USV^T$$

   where U and V are orthogonal matrices and S has non-zero values (the singular values) only on the diagonal.
4. Dimensionality is reduced by deleting all but the k largest elements of S, together with the corresponding columns in U and V. This truncation process is used to generate a k-dimensional vector space. Both terms and documents are represented by k-dimensional vectors in this vector space.
5. The similarity of any two objects represented in the space is reflected by the proximity of their representation vectors, generally using a cosine measure.

Extensive experimentation has shown that proximity of objects in this space is an effective surrogate for conceptual similarity in many applications [2].

3 Related Work

Numerous investigators have studied the effects of training data on LSI performance. This section summarizes key results from these studies, emphasizing three principal factors—quantity of training data, relevance of training data, and use of multiple LSI spaces.

3.1 Quantity of Training Data

In a seminal paper on LSI, Landauer and Dumais described synonym-matching experiments using questions from the Test of English as a Foreign Language (TOEFL). They analyzed LSI performance for six different sizes of training set, from 2,500 to 30,473 articles. They found an approximately logarithmic increase in discrimination capability with increasing training set size [3].

Using a collection of 364 news articles, Lee et al. compared human and LSI judgments of document similarity. They employed two training corpora, one consisting of 50 documents to be compared and one with 314 similar documents added. Using just the 50 comparison documents as the training corpus they obtained a correlation of approximately 0.54 with human similarity judgments. With the additional 314 training documents they obtained a correlation of 0.60, which was very close to the measured inter-rater correlation (0.605) for human evaluations of similarity of the documents [4].

Zelikovitz tested LSI categorization capabilities using five collections ranging from 1,000 to 2,472 documents. In each case she varied the amount of training data from 20 to 100 % of the training documents. In all five cases there was a continuous increase in categorization accuracy as the amount of training data increased [5].

Klebanov and Wiemar-Hastings used LSI as a source of world knowledge for pronominal anaphora resolution. Their test data was chosen from the Wall Street Journal. An initial test employing 15.8 MB of training data from the Wall Street Journal achieved a 27 % improvement over a baseline comparison. Reducing the amount of training data by 60 % had no impact on performance. Changing to training data from the Brown Corpus also had no significant impact [6].

Wiemer-Hastings et al. employed LSI in evaluating student answers to questions in a course on computer literacy. They employed four training sets, varying from 0.35 to 2.3 MB of text. In comparing LSI with human evaluations, they found that performance generally (but not strictly) increased with increasing training set size. Using mixtures of moderately related and highly related text, they found that either a rough balance of the two or a slight excess of highly specific texts yielded the best results [7].

In his thesis, Jung compared human and LSI-derived semantic similarity ratings of pairs of documents using a test set of 50 news articles. With an LSI space built from just the 50 test documents, the average correlation with human judgment was 0.46. Adding 314 similar news articles to the training set increased the average correlation to 0.54. Adding an additional 4,172 news articles increased the average correlation to 0.67 [8].

Bellegarda et al. examined the performance of LSI in the spam detection application of Apple e-mail in Mac OS 10.2. Once a level of 100 training e-mails was reached, there was little change when the amount of training data was tripled [9].

3.2 Relevance of Training Data

It has long been recognized that the training corpus used with LSI should match the task objective. As noted by Soto in 1999: "Since LSA learns about language exclusively from the training texts, it is very important to choose the right corpus for the specific situation to be modeled" [10].

Olmos et al. conducted tests of the efficacy of LSI for evaluating document summaries. Using an LSI space built from 390 summaries, they compared LSI-based measures with grades generated by four human judges Adding 63 training documents similar to the test topic had a small negative impact. Adding two million diverse training documents had a significant positive impact [11].

Kurby et al. conducted studies of LSI as an indication of reading strategies of students for scientific texts. They found that the correlation with human expert judgment was significantly better (0.67 vs. 0.55) when the LSI training corpus consisted of 273 science texts than when it was generated from thousands of general texts. Little change in performance was noted when this scientific corpus was reduced in size by 40% [12].

Kaur and Hornof tested LSI in predicting user click behavior for college websites. They experimented with three training sets—a generally-related collection consisting of 37,651 paragraphs, and two more-focused collections, consisting of 3,990 and 12,669 paragraphs, respectively. They observed a slight improvement in performance with the more-focused training data, but no significant difference between the two focused collections [13].

Bellegarda employed LSI as a component in statistical language modeling for speech recognition. In one experiment he used a test corpus of Wall Street Journal articles read aloud. Using 87,000 Wall Street Journal articles as a training corpus, his hybrid bigram-plus-LSI model achieved a 14% word error rate reduction in comparison to a standard bigram model. Using collections of 84,000–224,000 Associated Press news articles as training data yielded word error rate reductions of only 2–4% [14].

Perez et al. applied LSI to automated assessment of students' free-text answers to quizzes. They used two training sets; a focused one consisting of 1,929 student answers and a more general one composed of 142,580 articles on computer science. Their LSI-based assessment technique achieved somewhat higher correlation with human scores when the focused training corpus was used (0.43) than when the much larger general corpus was used (0.40) [15].

Zelikovitz and Hafner used Boolean searches to locate documents for use as background in enhancing LSI for categorization tasks. Using four sets of data ranging from 953 to 2,472 examples, they showed improved categorization results in three out of four cases [16]. In a variant of this work, Zelikovitz and Kogan used information gain to rank all of the words in the training set. They then retrieved background documents using queries based on the words from each class having the highest information gain. Incorporating the background data into the LSI training sets improved the categorization accuracy of physics article titles by 5 % and that of veterinary article titles by 53 % [17].

Zelikovitz and Marquez examined text categorization for three test sets, using background data selected based on TFIDF-weighted vectors of term occurrences. They conducted tests of 10 %, 20 %, up to 100 % of available sets of background documents. They found that, for larger collections, most of the gain in categorization accuracy that was obtained from using background data occurred when using only 50 % of the available background data. For smaller data collections, up to 75 % of the background data was needed in order to reach maximum performance [18].

Olde et al. studied the effects of different training corpora in comparing human and LSI-based evaluations of student responses to conceptual physics questions. They created a training corpus of 3,778 paragraphs of text from physics books. They then created four smaller corpora (416–3,445 paragraphs) by progressively eliminating peripherally relevant material, historical material, explanations of discarded theories, and discussions of incorrect chains of reasoning. Using all but the smallest collection, the comparisons between LSI-generated grades and human-assessed grades showed very little difference [19].

Mohler and Mihalcea studied the use of LSI for short answer grading for 21 questions in computer science. They measured Pearson correlation coefficients between human evaluations and LSI–generated evaluations of answers. They used five different training corpora based on lecture notes, documents from the British National Corpus (BNC), and Wikipedia. Performance was correlated with the size of the training set: a large general Wikipedia collection (1.8 GB) yielded better results than a smaller one (0.3 MB). A medium-sized (77 MB) CS-related Wikipedia collection also yielded better results than the lecture notes [20].

Haley et al. also analyzed use of LSI for grading of student answers to questions. They used a training space consisting of over 45,000 paragraphs of generally relevant text, plus variable numbers of student answers not included in the test set. They compared LSI-based evaluations and human evaluations for two questions. Surprisingly, the best results were obtained when only 50–80 of the 627 available answers were included in the training set [21]. In her thesis, however, Haley noted that, in similar testing, best results were obtained for other questions when using the maximum amount of available training data [22].

Cox and Pincombe conducted studies which compared LSI and human judgments of document similarity, using a collection of 50 newsmail articles. Building the LSI space from just the 50 documents whose similarity was to be judged yielded an LSI-human correlation of 0.54. When 314 similar articles were added to the training space, the correlation increased by 11 %. This value was not significantly different from the inter-rater correlation of the human judges (0.599 vs. 0.605). Somewhat surprisingly, adding another 2,432 documents from the same newsmail source (but a different time frame) reduced the correlation by 22 %. Creating the training set from 2,482 documents from a different source (the Hansard corpus of Canadian Parliament proceedings) yielded even lower correlations.

Cox and Pincombe also experimented with cross-lingual LSI. They used collections of 2,482 pairs of documents in English and French from both the Canadian Hansard and Amnesty International collections. Using mate matching as a similarity measure, they observed a considerable drop in match rate (0.87–0.63) when the training and test documents were drawn from different sources [23].

Stone et al. conducted experiments with LSI for similarity analysis of pairs of paragraphs. They used two test sets, one from the World Entertainment News Network (WENN) and the other from the Australian Broadcasting Corporation (ABC). For the WENN tests, where they employed 12,787 training and 23 test documents from the same corpus, they observed a correlation of 0.48 between the LSI and human evaluations of paragraph similarity. For the ABC tests, where the 55,021 training and 50 test documents came from different corpora (although both news articles) the correlation was a surprisingly low 0.12.

Stone and his collaborators then created focused corpora from Wikipedia, using Boolean queries automatically derived from the titles of the WENN documents and manually created from the ABC documents. Two groups of focused corpora were created; one group of 1,000 documents and the other of 10,000 documents. Each of these groups contained four training sets, one consisting of the entirety of each Wikipedia article, and the others consisting of the first 100, 200, or 300 words of each article. For both the WENN and ABC tests, the best performance was obtained using the first 100 words of 1,000 Wikipedia articles. For both test sets there was a dramatic improvement in performance as the portion of the Wikipedia articles used in the training was decreased. The authors suggested that this may in part be due to disambiguation—the initial portions of a document will tend to be focused, thus typically only one sense of each polysemous word will occur in them.

The authors also conducted a test of transductive learning—combining the 50 test documents of the ABC test with 1,000 Wikipedia articles as the training set. This yielded an 18% increase in correlation with human judgments. The resulting correlation (0.6) was close to the 0.605 observed for human inter-rater reliability for that data set [24].

Zelikovitz also tested LSI in a case where the test documents are available at the time the LSI space is built. For this transductive learning situation, incorporation of the test documents into the LSI training set yielded significant categorization improvement for all five of the test collections studied [5].

Zelikovitz and Hirsh studied a "second-order" transductive learning approach, where each background item was selected based on similarity to both a training example and a test document. The idea was that relations between words in such a background document might help to "bridge" context in the respective test and training documents. They obtained improvements in most cases, especially for tests with few training items and for those with short training items (3–9 words) [25].

3.3 Local LSI Spaces

Several relevant investigations have been carried out in which individual LSI spaces were generated for each category or query of interest in a given application. This approach has been referred to as *local LSI*.

Hull demonstrated improved performance on a categorization task for a collection of 1,399 documents through use of a separate LSI space for each of 219 categories. These local LSI spaces were based in part on vectors derived from an LSI index of the entire training set [26]. In [27] he showed that including similar non-relevant documents when creating the local spaces gave better results than including only relevant documents.

Wiener et al. used a similar local LSI approach for categorizing documents from the Reuters 22173 corpus. They built separate LSI spaces for individual topics and for clusters of documents related to similar topics. They identified terms related to topics and then used these terms as queries to select documents that were used to build the corresponding LSI spaces. They felt that such focused spaces might better represent small, localized effects, such as correlated use of infrequent terms. They achieved improvements over global LSI, particularly for topics with few training examples [28].

Schütze et al. addressed the routing problem using the Tipster corpus. For each query, they identified 2,000 relevant documents using a modified version of the Rocchio algorithm. They then created a local LSI space from these documents and applied statistical classifiers to the reduced-dimension representations. They achieved 10–15 % improvement in comparison to relevance feedback techniques [29].

In his thesis, Jiang applied the local LSI approach in both monolingual and cross-lingual tasks. He employed a standard vector space model to select documents for creation of the local LSI spaces. Using the AP 1990 collection and 50 TREC-4 ad hoc queries, he showed a 44 % improvement in average precision for local LSI over global LSI, with 50 documents chosen for LSI training for each query. This result was better than cases where 100 and 200 training documents were chosen for each query. He obtained similar results for monolingual French, German, and Japanese retrieval and for English–French, French–German, and English–Japanese cross-lingual retrieval [30].

Liu et al. applied relevancy weighting to documents used in creating local LSI spaces for classification. Their tests employed documents from the 25 categories of the Reuters 21578 test set that had the most training examples. They showed an advantage of relevance weighting over creating the local LSI spaces without weighting. [31]

Ding et al. conducted document categorization tests using a variant of the local LSI approach that they termed Local Relevancy Ladder-weighted LSI. They employed SVM to select the documents for creating the local LSI spaces. They achieved a 1.1 % improvement in F1 on a collection of documents from the largest 25 categories of the Reuters 21578 test set [32].

3.4 Summary

Overall, previous investigations of the effects of training set size and composition on LSI performance have demonstrated the following:

- Adding training data can have a significant effect on LSI performance across a range of tasks.
- As a general tendency, performance improves as the amount of additional training data increases.
- Performance improvement is greatest when the additional training data is well-matched to the problem being addressed.
- Creating separate LSI spaces for individual user interest topics can improve performance in tasks such as categorization.

A key aspect of the issue of adding training data that is left unresolved by the previous work is how to determine what constitutes well-matched training data. It is the objective of the present study to address this issue.

4 Rationale for Adding Training Data

There are five principal ways in which additional LSI training data might improve performance for a given text processing task:

1. New terms relevant to the task might be incorporated into the training space. Accordingly, newly acquired documents that contain those new terms generally will be assigned LSI vectors that more accurately represent their content.
2. The overall representational quality of the LSI space for the task may be improved. The additional documents provide additional contexts for terms. If the contexts match the problem, the improved term occurrence statistics may produce more effective representation vectors for those terms.
3. Relationships among task-relevant terms in the training documents could be reinforced.
4. Relationships between task-relevant and task-irrelevant terms in the initial set of training documents may be diluted. Some terms in the initial training documents that are not particularly relevant to the task may have co-occurred with other terms in an incidental fashion. If the incidental terms occur less often in the added training documents, the relationships between those terms and the relevant ones will tend to be weakened.
5. The effects of word sense ambiguity may be reduced. In the added documents, contexts of ambiguous terms may emphasize specific senses of those terms. This may allow some degree of disambiguation of those terms.

Based on the above considerations, it would appear to be desirable to select additional training data items that:

- Incorporate new relevant terms.
- Incorporate relevant contexts.
- Are focused in content (and thus minimize the number of senses of ambiguous words that appear in them).

5 Experimental Results

The experiments described here focused on cross-lingual LSI. We considered the problem of comparing documents in two languages: English and Persian. The evaluation metric chosen was mate matching. Although this is not a highly discriminating metric, the observed differences are large enough that it is quite adequate for the testing reported here. The use case emulated here is where English-language documents have been provided as examples of user interests and the task is to find relevant Persian-language documents in an incoming document stream.

5.1 Cross-Lingual LSI

In cross-lingual LSI, the relationships between terms in two languages are derived through exploitation of a parallel corpus for those two languages [33]. For English and Farsi, for example, a collection of document pairs is required, where one member of each pair is an English-language document and the other is its Farsi translation equivalent. For each pair, the documents are combined and treated as a single document. These combined documents are then processed in the manner described in Sect. 2. The result is an LSI space containing representation vectors for both English and Farsi terms. The closer together the representation vectors for a given English term and a given Farsi term, the more similar those terms are in meaning.

Representation vectors for documents not employed when building the space can be created in the resulting space through a process referred to as "folding-in" [1]. In this process, the new document is assigned a vector that is the weighted sum of the vectors of the terms that it contains. Since conceptually similar terms in the two languages lie close together, two documents that are similar conceptually end up with representation vectors that lie close together in the space, independent of their language. This characteristic of the space provides a basis for cross-lingual processing applications including document retrieval, clustering, and categorization. This approach can be applied to N languages simultaneously, so long as an N-way parallel corpus is available.

5.2 Experimental Data

Several types of documents were used as candidate training data in the experiments. These were chosen to provide a variety of conceptual content, as shown in Table 1.

Table 1 Candidate training documents

Data type	# of documents
Financial news articles	751,281
General news articles	3,725,925
Medline abstracts	438,723
Short technical items	221
UN documents	3,100
U.S. patents (whole)	314,069
U.S. patents (¶s)	10,450,926
Wikipedia	3,327,054
Total	19,011,299

Our test data consisted of Persian-language technical journal articles together with English-language abstracts from those articles. The journal articles were downloaded from http://www.journals4free.com/?fq=language:per. The test collection consisted of 504 articles, primarily on medical subjects. The English-language abstracts are short, typically consisting of a title plus from 2 to 10 sentences of text. These abstracts were compared to the Farsi bodies of the documents. In order to make the task more meaningful, the title and Farsi abstract were removed from the body documents. The authors' names and affiliations also were removed from both the abstracts and the bodies. These steps were intended to prevent correlations of titles, author names, etc. from unduly affecting the matching. The references also were removed, as many of those were in English, while the emphasis here was on matching Farsi documents.

5.3 Methodology

The intent of the effort was to emulate a situation where English-language user expressions of interest (the English-language abstracts used as exemplars) were compared to arriving Farsi-language documents. The abstracts were not employed in building the cross-lingual LSI spaces, so that term coverage effects could be studied. The Farsi body documents also were not employed in building the cross-lingual LSI spaces employed (except for a final transductive test).

The English-language abstracts and the Farsi body documents were folded into cross-lingual LSI spaces and compared using the cosine measure. The evaluation metric was the percentage of abstracts for which the closest document in the space was the corresponding Farsi body document.

Various subsets and combinations of the available candidate training data were employed as parallel corpora for creating the English-Persian cross-lingual LSI spaces tested. In the case of the UN documents, human-generated Farsi translations of the English-language documents were available. Farsi translations for all of the other candidate training documents were generated using the Bing® machine translation (MT) system, version 10.6. Cross-lingual LSI spaces were created from the English–Farsi pairs, using typical parameters for LSI spaces (deletion of numbers, a 618-word English stopword list, a 307-word Farsi stopword list, pruning of terms that did not occur at least twice, and 300 dimensions). A commercial LSI engine, Content Analyst® version 3.10, was used to create the spaces.

5.4 Training Data Relevance

The first test was intended to indicate the relative effectiveness of the different types of training data for this task. Approximately 3,000 documents[1] were chosen at random from six of the collections and used in creating six English-Farsi LSI spaces. The results of the mate matching test as carried out in each of these spaces are shown in Table 2.

The data demonstrate the great importance of context for LSI training data. Over 80 % of the test documents deal with biology and medicine. The Medline abstracts, which discuss similar topics, are the most closely matched candidate data type in terms of context. Employing those documents for training yielded a 40 % higher match rate than for any other training set.

In this test, similarity of contexts trumped term coverage. The Medline-based training data yielded a much higher match rate than the patents, even though the patent-based LSI spaces provided similar coverage of English-language terms and higher coverage of Farsi terms.

Using whole patents yielded much better results than splitting the patents by paragraph. This probably indicates that whole patents provide better overall topic matches to the Farsi journal articles than the individual paragraphs do.

Good conceptual match of the training set can help to compensate for variability in term choice and deficiencies in the MT system used to create the parallel corpus. The 69 % match using Medline is achieved in this case even though less than 20 %

Table 2 Mate match variation with training set

Data type	# of training documents	English term coverage (%)	Farsi term coverage (%)	Match result (%)
General news articles	3,034	59.7	19.0	48.8
UN documents	3,127	45.2	12.1	49.3
Medline abstracts	3,100	63.4	18.6	69.0
U.S. patents (whole)	3,100	61.2	20.5	48.8
U.S. patents (¶s)	65,995	60.8	20.4	38.1
Wikipedia	3,095	44.7	14.4	31.5

---

[1] The 65,995 patent paragraphs were extracted from the 3,100 whole patents. Elimination of some very short paragraphs yielded variations in term coverage.

of the Farsi terms present in the test documents appeared in the MT output used to create the cross-lingual LSI space.

5.5 Training Data Quantity

Table 3 shows the results when using much larger amounts of training data of the different types. For Medline and Wikipedia training, performance increased with additional training data. However, it is clear that quantity of training data is not the only consideration. For example, increasing the number of patents used in training by two orders of magnitude actually had a small negative impact on results. Term coverage also is not the only consideration. The highly relevant contexts of the Medline abstracts yielded a 77% higher match rate than the Wikipedia articles, even though the latter yielded similar English term coverage and higher Farsi term coverage.

Additional testing was conducted to provide a more granular indication of performance versus training set size. Figure 1 shows the variation in mate match percentage for collections of Medline, News, and Wikipedia articles varying from 3,000 to 1 million documents. In general, performance rises to a peak and then flattens out. This is true even though both English and Farsi term coverage are smoothly increasing as the sizes of the collections increase.

Combining different types of training data does not help in this situation. For example, Table 4 shows the results from combining Medline documents and UN documents with news articles.

In both of the combined training document tests, there is no synergy. Mixing Medline abstracts with news articles yielded performance intermediate between that produced by either document type alone. Mixing UN documents with news articles yielded performance worse than that for either type alone.

Table 3 Mate matching for larger training sets

Data type	# of training documents	English term coverage (%)	Farsi term coverage (%)	Match result (%)
General news articles	885,745	89.2	42.8	46.8
Medline abstracts	438,723	90.7	37.9	79.0
U.S. patents (whole)	310,450	84.5	37.7	48.7
U.S. patents (¶s)	3,158,856	83.8	36.1	36.1
Wikipedia	998,211	90.6	42.9	44.6

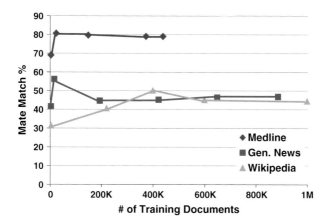

Fig. 1 Performance variation versus amount of training data

Table 4 Effects of mixing training data types

Data type(s)	# of training documents	English feature coverage (%)	Farsi feature coverage (%)	Match result (%)
News articles	15,936	72.5	25.9	56.2
UN documents	3,127	45.2	12.1	49.3
Medline abstracts	23,331	81.0	27.2	80.6
Medline + news	39,267	86.5	33.6	72.1
UN + news	19,063	73.3	29.3	32.3

5.6 Training Data Selection Technique

The Medline data collection happens to be particularly well-suited to this test case. For most applications, it is unlikely that such a well-matched collection of training data will be available. In addition, if ground truth is not available, there will, in general, be no effective means of telling if a given collection is well-matched or not.

These considerations provide strong motivation for finding a general method for selecting effective training data from available collections of documents, which does not require ground truth data for selection.

The dominant role of topic relevance demonstrated in the above test results motivated us to try using LSI itself as a mechanism for selecting additional training data.

Our proposed technique is to:

1. Accumulate a large and diverse collection of candidate training data.
2. Build an LSI representation space from all of the collected candidate training data.

3. Use the initially available data items (e.g., user-provided exemplars or queries) for the problem of interest as queries into this LSI space to select relevant additional training data.

The principal contributions of this paper are to demonstrate that:

1. The above-described technique is highly effective in cross-lingual LSI applications.
2. The technique can select appropriate training data from very large, heterogeneous collections of candidate training data.

This proposed technique was tested in the following manner:

- All of the available candidate training data was combined into one collection, comprising 19 million documents.
- An LSI space was created from that heterogeneous collection.
- Each of the English-language abstracts was employed as a query in that LSI space.
- For each of the abstracts, up to 500 additional training documents were chosen, using the criteria that the chosen training documents had to have a cosine of at least 0.6 in comparison to the abstract.[2]
- All of the training documents selected in this manner were translated into Farsi and the resulting document pairs were used to create a cross-lingual LSI space.
- The mate matching test was carried out in this cross-lingual space.

Using this technique, 84,002 documents were selected from the combined pool of candidate training documents (once duplicates were eliminated). In a cross-lingual LSI space built from these documents, a match percentage of 81.2 % was obtained.[3] This is better than any of the results previously obtained, including those from spaces built purely from Medline data.

This result shows that the proposed technique is capable of selecting highly effective training data even from large, heterogeneous collections of candidate training data, without requiring any ground truth data.

5.7 Local LSI

It was noted that 46 of the Farsi technical articles were from Electrical Engineering journals and the other 458 were from journals in the biomedical area. As a test of performance using a local LSI approach, we created two LSI spaces, one for the EE abstracts, and one for the biomedical abstracts, using the proposed technique for additional data selection. Combining the results from these two spaces yielded an

---

[2]Values varying between 50 and 5000 for number selected and from 0.5 to 0.7 for selection threshold were tried. The numbers 500 and 0.6 were found to be appropriate choices, but no attempt was made to find optimum values for these parameters.

[3]Randomly selecting an equivalent amount of training data from the combined collection and building a cross-lingual space using those documents yielded a match rate of only 51.4 %.

overall mate match rate of 89.7%. In an actual application, comparable knowledge of the topic distribution of user examples generally either would be available or could be determined through clustering.

For the local LSI case we ran a final test using transductive training. Including the Farsi body documents into the training set increased the size of the training set by less than 1%, but doubled the Farsi term coverage (from 34.8 to 68.4%). (The term coverage is not 100% primarily because of the pruning of terms that did not occur at least twice when creating the LSI space). In this transductive scenario, the match rate increased to 96.2%, showing the value of increased term coverage, so long as the contexts of the cross-lingual training documents are correct.

6 Conclusion

At least for the test case employed here, the degree of conceptual match of the training set was by far the most important factor in determining cross-lingual LSI performance. It is thus highly desirable to be able to choose appropriate training data for a particular application. We have presented a principled approach for automated selection of effective training data. This approach has two key features:

1. It is not dependent on availability of any ground truth data.
2. It is capable of selecting highly effective training sets from large, heterogeneous collections of candidate training data.

Figure 2 summarizes the results obtained using the proposed technique.

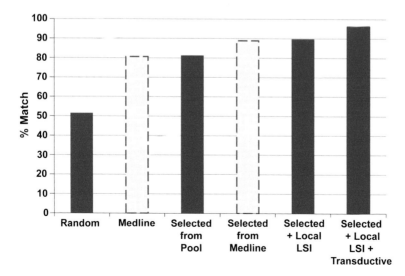

Fig. 2 Performance versus training data source

Principled selection of training data using the proposed technique provided a dramatic improvement over random selection. This result was further improved by employing two local LSI spaces to match the exemplar topic distribution. Using transductive training further increased performance. The Medline results in Fig. 2 are deemphasized (presented in dashed lines), because typically such a very-well-matched training set would not be available in an actual application.

The effectiveness of the approach was shown here for a cross-lingual document comparison application. However, the technique is generally applicable to LSI clustering, search, and discovery applications. Moreover, an analogous technique could be employed in situations involving LSI processing of other types of data such as audio, video, signals data, or images.

References

1. Deerwester, S., et al.: Improving information retrieval with latent semantic indexing. In: Proceedings of ASIS'88, Atlanta, GA, pp. 36–40 (1988)
2. Bradford, R.: Comparability of LSI and human judgment in text analysis tasks. In: Proceedings of Applied Computing Conference, Athens, Greece, pp. 359–366 (2009)
3. Landauer, T., Dumais, S.: A solution to Plato's problem: the latent semantic analysis theory of acquisition, induction, and representation of knowledge. Psychol. Rev. **104**, 211–240 (1997)
4. Lee, M., Pincombe, B., Welsh, M.: An empirical evaluation of models of text document similarity. In: Proceedings of 27th Annual Conference of the Cognitive Science Society, pp. 1254–1259 (2005)
5. Zelikovitz, S.: Transductive LSI for short text classification problems. Int. J. Pattern Recognit. Artif. Intell. **19**(2), 143–163 (2005)
6. Klebanov, B., Wiemer-Hastings, P.: The role of wor(l)d knowledge in pronominal anaphora resolution. In: Proceedings of International Symposium on Reference Resolution for Natural Language Processing, Alicante, Spain, 3–4 June 2002
7. Wiemer-Hastings, P., Wiemer-Hastings, K., Graesser, A.: Open Learning Environments: New Computational Technologies to Support Learning, Exploration and Collaboration. IOS Press, pp. 535–542 (1999)
8. Jung, K.: Mismatches between humans and latent semantic analysis in document similarity judgments. Doctoral thesis. University of New Mexico, July 2013
9. Bellegarda, J., Naik, D., Silverman, K.: Automatic junk e-mail filtering based on latent content. Autom. Speech Recognit. Underst. 465–470 (2003)
10. Soto, R.: Learning and performing by exploration: label quality measured by latent semantic analysis. In: CHI'99, pp. 418–425 (1999)
11. Olmos, R., et al.: An analysis of size and specificity of corpora in the assessment of summaries using LSA: a comparative study between LSA and human raters. Revista Signos (in Spanish) **42**(69), 71–81 (2009)
12. Kurby, C., et al.: Computerizing reading training: evaluation of a latent semantic analysis space for science text. Behav. Res. Methods Instrum. Comput. **33**(2), 244–250 (2003)
13. Kaur, I., Hornof, A.: A comparison of LSA, WordNet, and PMI_IR for predicting user click behavior. In: Proceedings of CHI 2005, Portland, Oregon, 2–7 April 2005
14. Bellegarda, J.: Exploiting latent semantic information in statistical language modeling. Proc. IEEE **88**(8), 1279–1296 (2000)
15. Perez, D., et al.: Automatic assessment of students' free-text answers underpinned by the combination of a BLEU-inspired algorithm and latent semantic analysis. In: Proceedings of 18th International Florida Artificial Intelligence Research Society Conference, FLAIRS (2005)

16. Zelikovitz, S., Hafner, R.: Automatic generation of background text to aid classification. In: Proceedings of FLAIRS Conference (2004)
17. Zelikovitz, S., Kogan, M.: Using web searches on important words to create background sets for LSI classification. In: Proceedings of FLAIRS Conference'06, pp. 598–603 (2006)
18. Zelikovitz, S., Marquez, F.: Evaluation of background knowledge for latent semantic indexing classification. In: Proceedings of Eighteenth International Florida Artificial Intelligence Research Society Conference, Clearwater Beach, Florida, USA (2005)
19. Olde, B., et al.: The right stuff: do you need to sanitize your corpus when using latent semantic analysis? In: Proceedings of the 24th Annual Meeting of the Cognitive Science Society, pp. 708–713 (2002)
20. Mohler, M., Mihalcea, R.: Text-to-text similarity for automatic short answer grading. In: Proceedings of 12th Conference of the European Chapter of the ACL, Athens, Greece, 30 March–3 April 2009, pp. 567–575
21. Haley, D., et al.: Measuring improvement in latent semantic analysis-based marking systems: using a computer to mark questions about HTML. In: Proceedings of Ninth Australasian Computing Conference, Ballarat, Victoria, Australia (2007)
22. Haley, D.: Applying latent semantic analysis to computer assisted assessment in the computer science domain: a framework, a tool, and an evaluation. Doctoral thesis, The Open University (2009)
23. Cox, W., Pincombe, B.: Cross-lingual latent semantic analysis. ANZIAM J. **48**, C1054–C1074 (2008)
24. Stone, B., Dennis, S., Kwantes, P.: Comparing methods for single paragraph similarity analysis. Top. Cogn. Sci. **3**, 92–112 (2011)
25. Zelikovitz, S., Hirsh, H.: Improving short-text classification using unlabeled background knowledge to assess document similarity. In: Proceedings of the 17th International Conference on Machine Learning, pp. 1183–1190
26. Hull, D.: Improving text retrieval for the routing problem using latent semantic indexing. In: Proceedings of 17th ACM SIGIR Conference, pp. 282–291
27. Hull, D.: Information retrieval using statistical classification. Doctoral thesis, Stanford University, November 1994
28. Wiener, E., Pederson, J., Weigend, A.: A neural network approach to topic spotting. In: Proceedings of Fourth Annual Symposium on Document Analysis and Information Retrieval (SDAIR'95), Las Vegas, NV, 24–26 April 1995
29. Schutze, H., Hull, D., Pedersen, J.: A comparison of classifiers and document representations for the routing problem. In: Proceedings of SIGIR, vol. 95, pp. 229–237 (1995)
30. Jiang, F.: Matrix computations for query expansion in information retrieval. Doctoral thesis, Duke University, September 2000
31. Liu, T., et al.: Improving text classification using local latent semantic indexing. In: Proceedings of ICDM'04, pp. 162–169
32. Ding, W.: LRLW-LSI: an improved latent semantic indexing (LSI) text classifier. In: Proceedings of Third International Conference on Rough Sets and Knowledge Technology, Chengdu, China, pp. 483–490
33. Dumais, S., Letsche, T., Littman, M., Landauer, T.: Automatic cross-language retrieval using latent semantic indexing. In: Proceedings of AAAI-97 Spring Symposium Series: Cross-Language Text and Speech Retrieval, 24–26 March 1997, pp. 18–24

Depth-First Heuristic Search for Software Model Checking

Jun Maeoka, Yoshinori Tanabe and Fuyuki Ishikawa

Abstract Software model checkers, such as Java PathFinder (JPF), can be used to detect failures in software. However, the state space explosion is a serious problem because the size of the state space of complex software is very large. Various heuristic search algorithms, which explore the state space in the order of the given priority function, have been proposed to solve this problem. However, they are not compatible with linear temporal logic (LTL) verification algorithms. This paper proposes an algorithm called *depth-first heuristic search* (DFHS), which performs depth-first search but backtracks at states that unlikely lead to an error. The likelihood is evaluated using *cut-off functions* defined by the user. We enhanced the search engine of JPF to implement DFHS and LTL search. Experimental results show that DFHS performs better than current algorithms for both safety and LTL properties of programs in many cases.

Keywords Software model checking · Heuristic search · JPF

J. Maeoka (✉)
Research and Development Group, Hitachi, Ltd., 292 Yoshida-cho, Totsuka-ku,
Yokohama-shi, Kanagawa 244-0817, Japan
e-mail: jun.maeoka.hf@hitachi.com

J. Maeoka · F. Ishikawa
The University of Electro-Communications, 1-5-1 Chofugaoka,
Chofu-shi, Tokyo 182-8585, Japan

Y. Tanabe (✉)
Tsurumi University, 2-1-3 Tsurumi, Tsurumi-ku,
Yokohama-shi, Kanagawa 230-8501, Japan
e-mail: tanabe-y@tsurumi-u.ac.jp

F. Ishikawa (✉)
National Institute of Informatics, 2-1-2 Hitotsubashi,
Chiyoda-ku, Tokyo 101-8430, Japan
e-mail: f-ishikawa@nii.ac.jp

© Springer International Publishing Switzerland 2016
R. Lee (ed.), *Computer and Information Science 2015*,
Studies in Computational Intelligence 614, DOI 10.1007/978-3-319-23467-0_6

1 Introduction

Model checking is a technology to verify the properties of models by exploring all possible states and execution paths of the models. Model checking was originally used for verifying hardware, software design, and communication protocols [12, 14]. In recent years, this technology has been applied to software source code and byte code. In this case, the size of the state space is typically very large, and it is often difficult to explore all the state space in a reasonable time frame. It can easily take months or years to reproduce the problem depending on the scale of the software. The technology is still useful in finding failures because the number of states until the first bug is found is typically much less than that of the entire state space. These failures often depend on the order of thread scheduling and are difficult to find by testing. In model checking, the model checker controls the scheduling and eventually checks every possibility, while it is not controlled at all in testing.

To find bugs faster, it is important to search parts of the state space whose possibility of containing bugs is higher than others. One approach is heuristic search [20], which is basically an idea of prioritizing the candidates to find the best goal or quickly find a goal. Heuristic search algorithms can be applied to software model checking to solve the state explosion problem described above. Various search algorithms and priority functions have been proposed [8, 9, 11, 18].

In the context in which the entire space is too large to explore, it is meaningless to guarantee that algorithms will explore all states. Rather, it is important for algorithms to quickly find the first bug. Our algorithm does not logically guarantee the absence of failures when it does not find failures. However, experimental results show that it reaches the first bug in a reasonable time frame, often faster than current algorithms.

There are two types of properties we are interested in. The first is called a safety property, which can be checked by visiting a state. The second is called a linear temporal logic (LTL) property, which requires more complex algorithms based on depth-first search (DFS). Traditional heuristic search algorithms are not compatible with LTL algorithms. Studies [8, 9, 11, 18] have been conducted only for safety properties.

We propose a different heuristic search algorithm based on DFS called depth-first heuristic search (DFHS). Our algorithm performs DFS but backtracks at states that less likely lead to error states. The likelihood is evaluated using cut-off functions defined by users. The DFHS algorithm can be applied to verification of LTL properties. It can also be applied to safety properties.

For safety properties, DFHS explores the space in a different manner from current heuristic search algorithms. Therefore, it is expected that several bugs can be found earlier. In fact, experiments showed that DFHS performs better in many cases.

We implemented DFHS in Java PathFinder (JPF), a software model checker for Java.[1] Because JPF originally does not support LTL verification, we add a search engine for LTL verification. We evaluated DFHS for safety and LTL verification.

---

[1]Java is a registered trademark of Oracle and/or its affiliates. Other names may be trademarks of their respective owners.

The rest of the paper is organized as follows. We first introduce related work in Sect. 2 and current algorithms in Sect. 3. We describe DFHS in Sect. 4. We explain the implementation of DFHS in JPF in Sect. 5 and the experimental results in Sect. 6. We conclude the paper in Sect. 7.

2 Related Work

Heuristic search is a technology for efficient search in artificial intelligence [20]. Heuristic search algorithms, such as Best-First, A*, and Beam, control the search order of states by calculating the priority score of each state using heuristic functions. Heuristic search is applicable to model checking to solve the state explosion problem [8, 9, 11, 18]. Edelkamp et al. proposed HSF-SPIN [8], which uses heuristic search with SPIN [12]. They argue that it can find shorter counterexamples than traditional search. Groce et al. used heuristic search in JPF [11]. They introduced heuristic functions based on the number of branch selections and number of thread interleavings. The heuristic search algorithm of Rungta et al. [18] uses warning information of a software static analyzer to estimate the distance from the current state to an error state.

There are other approaches for quickly finding bugs [15, 16]. The algorithm proposed by Parízek et al. [16] is based on DFS and randomly backtracks. The probability corresponds to the search depth. They showed better experimental results than those of current heuristic search algorithms. Context-bounded search by Musuvathi et al. [15] limits the number of context switches. It quickly finds errors compared to traditional search algorithms. These algorithms can be regarded as special cases of our algorithm.

For verifying LTL properties, an automata-based approach is frequently used [22]. With this approach, the target space to be explored is the synchronized product of the state space and the Büchi automaton converted from an LTL formula. Typically, the target space is explored using algorithms based on DFS. Nested depth-first search (NDFS) is one of the most well known algorithms [3], and algorithms that use strongly connected components (SCCs) have also been proposed [4]. In parallel search context, algorithms based on breadth-first search are also used [1].

Sun et al. proposed an SCC-based algorithm for LTL verification with fairness [21]. The algorithm walks through the entire state space with DFS to find SCCs. Their paper also introduces efficient enhancement for fairness consideration. The algorithm checks the compliance to fairness conditions of counterexamples after they are found. Our cut-off approach might be applied to this SCC-based algorithm in future work.

For our study, we applied our algorithm to the most well known LTL verification algorithm. However, various efficient LTL verification algorithms have been proposed [5, 10] and our algorithm might be applicable to them in the future.

Several extensions of JPF for LTL verification using the automata-based approach have been reported [6, 13]. Lomabardi's implementation [13] provides a way to specify LTL formulas as annotations in the source code and explores the target space with NDFS. Cuong et al. [6] concentrated on method calls to realize a light-weight verification. Unlike these implementations, our implementation supports atomic propositions that are used to express fairness conditions. Algorithms for verification with fairness conditions have been studied. In particular, for strong fairness conditions, transition-based automata have been proposed for efficient exploration [7]. We take a different approach since we concentrate on efficient error detection rather than verification.

In terms of reducing memory usage, bit state hashing and state space caching are well known approaches [10, 17]. These are applicable to traditional heuristic algorithms and ours independently with the search algorithms discussed in this paper. The JPF applies bit state hashing.

3 Current Algorithms

In this section, we first explain DFS and NDFS. We then introduce best-first search (BFS).

3.1 Depth-First Search

Depth-first search is one of the most common search algorithms. Figure 1 shows how DFS searches a goal. As this figure shows, DFS goes forward toward a successor of the current node until it reaches an end node or a visited node, in which case it backtracks.

The JPF, a software model checker we used for this study, uses DFS as a default search algorithm. In terms of software model checking, a state is a state of a running program, and the goal is the state containing an error.

Fig. 1 Depth-first search (DFS)

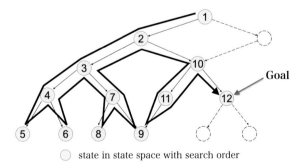

state in state space with search order

3.2 Nested Depth-First Search

Nested depth-first search [3] is a well known algorithm for LTL verification. It is used to explore a synchronized product of a given state space and the Büchi automaton converted from a given LTL formula. It consists of two DFS procedures called blue DFS and red DFS. Blue DFS is the main loop that searches for accept states recursively. Before backtracking from an accept state, blue DFS suspends searching and red DFS starts searching from the state to find a state on the current path of blue DFS. If such a state is found, NDFS reports the current path as a counterexample.

3.3 Best-First Search

There are various heuristic search algorithms for software model checking. In this section, we introduce BFS, which has been reported as the most efficient algorithm [11]. A search engine of BFS is included in JPF.

Figure 2 gives an overview of BFS. For the transition system shown in the figure, BFS reaches the goal faster than DFS and attempts important nodes first. The word "important" means "likely to lead to a goal faster". This importance is expressed by a function called a heuristic function.

Best-first search manages search nodes in the priority queue. BFS takes the best node out of the queue and puts all its successors into that queue. The nodes in the queue are sorted by their priority values determined by the heuristic function. This process continues until all nodes are checked or a goal is found.

Figure 3 is the pseudo code of BFS. Here, h() is the heuristic function, which returns the priority value, Queue is the priority queue sorted by the priority value, remove_best() pops a node having the highest priority from the queue, and insert() inserts a new node into the queue at the appropriate location. From the implementation view point, the size of the queue is often limited because of an

Fig. 2 Best-first search (BFS)

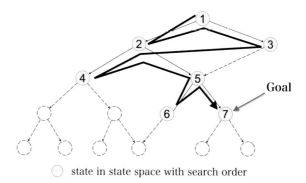

Goal

state in state space with search order

```
best_first_search:
  sorted_queue Queue = [S0]
  set Visited = {S0}
  while (S = remove_best(Queue)) != null
      while (S' = get_next_successor(S)) != null
          if is_error_state(S')
              report_error(S')
              return
          else if not exists(Visited, S')
              Hval = h(S')
              insert(Queue, S', Hval)
              add(Visited, S')
```

Fig. 3 BFS algorithm

implementation limitation. In such cases, nodes with lower priorities are dropped from the queue and their successors can be ignored.

When BFS is applied to software model checking, heuristic functions can be calculated using the structure of the state space (e.g. the depth of the node), conditions of threads (e.g. the number of blocked threads), and information from the static analyzer (e.g. the distance from the location of assertion statements). For example, the heuristic functions from [11] are as follows.

Interleaving: The fewer previously selected thread appear in recent n transitions, the higher the heuristic function scores.
MostBlocked: The higher the number of threads that are blocked, the higher the heuristic function scores.
Random: The heuristic function randomly scores.

4 Proposed Algorithm

The traditional heuristic search algorithms described in the previous section exhibit the problem mentioned in Sect. 1. We propose DFHS, which can be applied not only to safety properties but also LTL properties. The DFHS algorithm modifies DFS and NDFS with the following three features.

4.1 Cut-Off Function

The first feature is a cut-off function. The DFHS algorithm cuts off further search from states that are unlikely to lead to errors. The likelihood is expressed by a function called a cut-off function that returns true or false for a given state. When the value of the cut-off function at the current state is true, DFHS backtracks. A cut-off function f can be automatically defined from a heuristic function h of a traditional heuristic search algorithm and a threshold t: if $h(s) > t$ then $f(s) =$ true; otherwise, $f(s) =$ false. However, the cut-off functions we used in the experiment and describe in Sect. 6 do not fall into this category.

The following are examples of cut-off functions.

Interleaving(Int):   backtracks if the current thread had been selected in the recent past.

NonConsecutive(NonCon):   backtracks if a thread is selected twice consecutively.

LessInterleaving(LessInt): backtracks if the number of interleavings from the beginning exceeds a threshold.

Blocked: backtracks if the number of blocked threads had not increased in the recent past.

Random: backtracks randomly with specified probability.

The DFHS algorithm uses cut-off functions in both blue DFS and red DFS in NDFS.[2]

4.2 Successor Order

The second feature is successor order. Because DFHS is based on DFS, it is impossible to apply the idea of BFS globally. However, we can locally apply it by defining the search order of the successors of a state. The DFHS algorithm visits them in user-specified order.

The following are examples of successor orders.

Default: prefers a thread with small thread id.
Interleave(Int):   prefers a thread other than the current thread.
LessInterleave(LessInt):   prefers the current thread.
Random: random order.

4.3 Fairness Condition Optimization

The third feature only applies to LTL properties under fairness conditions. It is well known that in software model checking, meaningless counterexamples are often reported with unfair thread scheduling. Fairness conditions exclude such unfair scheduling. This can be achieved naively by modifying a given LTL formula. However, this approach is not very efficient because the modified LTL formula is often

---

[2]The correctness of NDFS depends on the fact that blue DFS visits all children of a state before red DFS starts for the state. By applying a cut-off function to NDFS, DFHS does not guarantee this condition. However, a reported counterexample with DFHS is still correct, which is sufficient for our purpose to find errors.

long, which makes the state space huge. With DFHS, we explore the state space without modifying the LTL formula, inspired by [21].

- When a counterexample is found, we check whether it satisfies the fairness condition. If not, we continue to find other counterexamples.
- We continue until a fair counterexample is found.

Note that DFHS may overlook fair counterexamples even when they exist. Therefore, it cannot be used in a verification context. In most cases, however, DFHS quickly finds fair counterexamples, as we discussed in Sect. 6.

Figure 4 shows a pseudo code of the DFHS algorithm. Here, `cutoff_function()` determines whether it should keep searching forward. The cutoff function is called only at the deeper states than a certain depth in order not to cut off states too much. The term `get_next_successor_in_order()` returns the successors of the current state individually in the order it prioritizes.

Figure 5 shows how DFHS works. It checks the value of the cut-off function each time it reaches a new state. If the value is true, DFHS backtracks. In this figure, it backtracks at the state labeled 3.

```
depth_first_heuristic_search:
  stack Stack = [S0]
  set Visited = {S0}
  while (S = top(Stack)) != null
    S' = get_next_successor_in_order(S)
    if S' == null
      pop(Stack)
    else if is_error_state(S')
      report_error(S')
      return
    else if not exists(Visited, S')
      add(Visited, S')
      if size_of(Stack)>predefined_depth_limit
        if not cutoff_function(S')
          push(Stack,S')
```

Fig. 4 Depth-first heuristic search (DFHS) algorithm

Fig. 5 DFHS

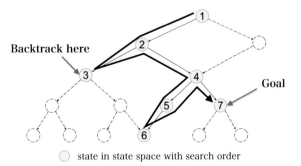

state in state space with search order

5 Implementation

In this section, we explain the implementation of DFHS. We used JPF [23] to implement it. First, we describe the verification mechanism of JPF then explain the enhancement of JPF to support DFHS and NDFS.

5.1 Java PathFinder

The JPF has its own Java virtual machine and executes a Java program to detect errors by visiting all possible program states. A state consists of the current locations of all threads, values of all variables and fields of objects, statuses of synchronization locks, and so on. In other words, a program state is a snapshot of the running program at each execution point.

The most important non-deterministic factor is thread interleaving. If more than one thread is runnable, JPF picks up one and proceeds with the execution. The default search algorithm is DFS, and heuristic search is available.

5.2 Enhancement of JPF

The JPF consists of two parts, a Java virtual machine and configurable search engine. The search engine can be modified or replaced by users. Event listeners can be added to the search engine, assigned to a particular type of event, and called when such an event occurs.

We enhanced JPF with these mechanisms to support NDFS and DFHS. Figure 6 shows the original architecture of JPF and our enhancements. We explain these enhancements in detail below.

5.2.1 Nested Depth-First Search

The JPF does not originally support LTL verification, but there are several implementations that support LTL verification [6, 13]. We also developed a search engine for JPF that implements LTL verification. The search engine performs NDFS [3] in the synchronized product of the original state space and the Büchi automaton corresponding to a given LTL formula.

We developed a class for atomic propositions and a method for modifying their truth values. The user can maintain them in the listeners. We also developed a mechanism that breaks the current transition when the value of an atomic proposition has changed. The atomic propositions that express that a thread is enabled and that a thread has just run are provided in order to express fairness conditions.

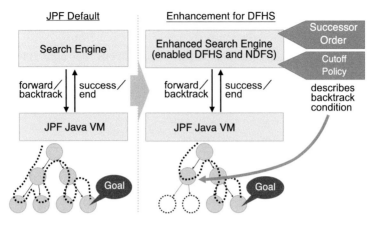

Fig. 6 Enhancement of JPF

For example, if the liveness property of the request-response with a strong fairness condition is to be tested, the user can specify the following property in a configuration file:

```
ltl.formula =
    ([]<>th_enb1->[]<>th_run1) ->
    ([]<>th_enb2->[]<>th_run2) ->
        [](p -> <> q)
```

and declare that `th_enb`*i* and `th_run`*i* express that thread *i* is enabled and has just run, respectively.

5.2.2 DFHS

We developed a generic listener class for cut-off functions. The listener evaluates the value of such functions. If the value is true, it sets a flag to indicate that the search engine should backtrack at the current state. The listener is called every time JPF reaches a state. The user extends the generic listener class to make a cut-off function possible. Figure 7 shows an example of a cut-off function that commands the search engine to backtrack if a thread runs three consecutive times.

The JPF has a mechanism called a choice generator that individually returns the enabled threads at a state. The choice generator used in DFS returns threads in the order of the thread id. We implemented a method that reorders the threads according to the order that the user specifies.

To make the feature explained in Sect. 4.3 possible, we also implemented optimization for strong fairness conditions. Our enhanced search engine stores the values of atomic propositions, including those that express a thread that is enabled or has just run, at each state in the current search path. When a counterexample is found,

Fig. 7 Implementation of
cut-off function

```
boolean check(BASearch search, SPNode spn) {
    // threadIdHistory: selected threads
    int size = threadIdHistory.size();
    int t1 = threadIdHistory.get(size - 1);
    int t2 = threadIdHistory.get(size - 2);
    int t3 = threadIdHistory.get(size - 3);
    if (t1 == t2 && t2 == t3) {
        return true; // backtrack
    }
    return false; // continue
}
```

the search engine checks if the counterexample conforms to the strong fairness condition, i.e., if each thread is either always disabled or runs at least once in the loop part of the counterexample.

6 Experiment

We evaluated DFHS for safety and LTL properties with benchmark programs.

6.1 Safety

We used all programs in a Java benchmark suite [19] that are publicly available from [2]. This benchmark suite is designed for evaluating a verification tool and contains bugs of multi-threaded programs such as deadlocks, uncaught exceptions, or assertion violations.

We applied the heuristic functions described in Sect. 3.3, cut-off functions in Sect. 4.1, and successor order in Sect. 4.2. The cut-off functions are not called if the depth is less than 5. We set the memory limit to 2,048 MB and timeout to 15 min. Performance was measured based on the sum of created and visited states until an error was detected.

Table 1 lists the results of DFS and BFS. The number in each cell represents the number of states when the model checker successfully found an error, and the elapsed time is shown in parentheses; formatted as mm:ss.[3] The symbol "−" denotes failure to find errors, OOM denotes out-of-memory, and TO denotes time-out. The best result of each algorithm is boldfaced. Table 2 lists the results of DFHS. The best case of each program in Table 1 is in the leftmost column. For safety properties, we applied two features, cut-off functions and successor orders. At the second line of the title row, (1) means the application of cut-off functions and (2) means that of

---

[3] The number of states is related to elapsed time and reproducible while the elapsed time differs at every executions. Therefore we omit the elapsed time field in the following tables to save space.

Table 1 Safety results (current algorithms)

Program	DFS	BFS		Random
		Interleaving	MostBlocked	
Account	**636 (00:01)**	603,334 (01:14)	25,282 (00:05)	5,056 (00:02)
AccountSubtype	TO	TO	**966,371 (02:33)**	5,250,146 (14:55)
Airline 10 3	TO	**287,121 (00:29)**	**287,121 (00:30)**	–
AlarmClock	**544 (00:01)**	55,410 (00:10)	2,558 (00:02)	8,587 (00:03)
AllocateVector	TO	194,800 (00:27)	1,024,422 (02:05)	**4,098 (00:03)**
Apps	2,598,794 (06:24)	–	**84,780 (00:19)**	–
BoundedBuffer	**2,658 (00:02)**	430,838 (01:02)	176,495 (00:25)	TO
Clean 3 3 3	**98 (00:01)**	347,208 (00:48)	532,789 (01:47)	1,746,836 (04:42)
Daisy	99,915 (00:21)	6,747 (00:04)	**2,799 (00:03)**	23,048 (00:10)
Deadlock1	78 (00:00)	198 (00:00)	206 (00:00)	**67 (00:00)**
Deadlock2	**12 (00:00)**	35 (00:00)	35 (00:00)	39 (00:00)
DEOS	87,213 (05:56)	34,775 (01:57)	34,775 (01:55)	**10,513 (01:51)**
DiningPhil 10	282 (00:02)	195,361 (06:45)	**194 (00:00)**	2,656 (00:02)
Piper	**25,464 (00:06)**	TO	TO	2,804,790 (10:02)
LinkedList	**3,146 (00:03)**	107,000 (00:22)	TO	25,159 (00:08)
LoseNotify 3 3	**14 (00:00)**	11,792 (00:05)	437 (00:01)	61 (00:01)
NestedMonitor	**5 (00:00)**	19 (00:00)	11 (00:00)	20 (00:00)
ProCon	**13,530 (00:04)**	763,092 (01:51)	3,177,820 (06:21)	1,624,666 (03:35)

(continued)

Table 1 (continued)

Program	DFS	BFS		Random
		Interleaving	MostBlocked	
Raxextended	TO	78,577 (00:14)	94,399 (00:19)	**1,288 (00:01)**
RaxEnvFirst	OOM	TO	TO	**341,431 (01:08)**
Reorder 5 2	297,683 (00:33)	21,487 (00:06)	12,434 (00:04)	**2,891 (00:02)**
ReadersWriters	2,535,301 (04:38)	49,309 (00:12)	17,925 (00:08)	**3,341 (00:03)**
SleepingBarber	**17 (00:00)**	549 (00:01)	108 (00:00)	201 (00:00)
TwoStage 5 5	TO	–	–	**90,321 (00:15)**
Wronglock 10 1	17,480 (00:06)	202,439 (00:31)	**459 (00:01)**	2,612 (00:02)

Table 2 Safety Results (DFHS)

Program	Best of Table 1	DFHS (2) DFS/Int	(1) Int	NonCon	LessInt	Blocked	Random	(1)+(2) Int/Int	NonCon/Int	LessInt/LessInt	Random/Random
Account	636 (00:01)	609 (00:01)	343 (00:01)	386 (00:01)	636 (00:01)	20,652 (00:06)	561 (00:01)	312 (00:01)	**234** (00:01)	673 (00:01)	529 (00:01)
Account Subtype	966,371 (02:33)	TO	**8,036** (00:03)	TO	TO	TO	TO	TO	TO	TO	TO
Airline 10 3	287,121 (00:29)	TO	1,921,393 (02:42)	TO	TO	TO	TO	**632** (00:01)	TO	5,181 (00:03)	TO
Alarm Clock	544 (00:01)	328 (00:01)	**60** (00:00)	1,140 (00:01)	544 (00:01)	544 (00:01)	–	208 (00:01)	88 (00:01)	612 (00:01)	–
Allocate Vector	4,098 (00:03)	**35** (00:00)	TO	55 (00:01)	TO	TO	220 (00:01)	**35** (00:01)	**35** (00:01)	TO	39,888 (00:09)
Apps	84,780 (00:19)	TO	**499** (00:01)	3,512,404 (07:59)	–	TO	1,224,545 (02:55)	3,088,261 (07:35)	TO	–	125,107 (00:21)
Bounded Buffer	2,658 (00:02)	3,885 (00:02)	TO	30,149 (00:17)	473,986 (01:02)	674,175 (01:26)	**182** (00:01)	TO	3,885 (00:04)	22,454 (00:08)	501 (00:01)
Clean 3 3 3	**98** (00:01)	131 (00:00)	1,754 (00:02)	234 (00:01)	TO	**98** (00:00)	270 (00:00)	390 (00:01)	131 (00:01)	TO	26,084 (00:05)
Daisy	**2,799** (00:03)	98,125 (00:21)	99,915 (00:22)	–	99,915 (00:25)	4,778 (00:04)	–	98,125 (00:20)	–	99,434 (00:24)	–
Deadlock -1	67 (00:00)	54 (00:00)	78 (00:00)	75 (00:00)	78 (00:00)	78 (00:00)	–	54 (00:00)	52 (00:00)	100 (00:00)	**24** (00:00)
Deadlock -2	12 (00:00)	11 (00:00)	12 (00:00)	12 (00:00)	12 (00:00)	12 (00:00)	12 (00:00)	11 (00:00)	11 (00:00)	23 (00:00)	**9** (00:00)
DEOS	**10,513** (01:51)	87,213 (05:55)	87,213 (06:24)	87,213 (06:24)	87,213 (06:31)	87,213 (06:16)	–	87,213 (06:23)	87,213 (06:24)	87,213 (06:13)	–
Dining Phil 10	194 (00:00)	348 (00:01)	294 (00:01)	332 (00:02)	TO	262 (00:01)	268 (00:01)	483 (00:01)	348 (00:01)	TO	**118** (00:00)
Piper	25,464 (00:06)	25,186 (00:06)	TO	26,969 (00:07)	TO	TO	4,253 (00:02)	103,642 (00:16)	25,186 (00:07)	TO	**1,446** (00:01)
Linked List	3,146 (00:03)	784 (00:01)	3,146 (00:02)	521 (00:01)	3,146 (00:03)	**404** (00:01)	–	784 (00:01)	20,899 (00:05)	3,278 (00:02)	–
LoseNotify 3 3 3	**14** (00:00)	19 (00:00)	2,471 (00:02)	26 (00:00)	**14** (00:00)	**14** (00:00)	30 (00:00)	27 (00:00)	19 (00:00)	**14** (00:00)	35 (00:00)
Nested Monitor	**5** (00:00)	13 (00:00)	**5** (00:00)	**5** (00:00)	**5** (00:00)	**5** (00:00)	**5** (00:00)	13 (00:00)	13 (00:00)	**5** (00:00)	11 (00:00)

(continued)

Table 2 (continued)

Program	Best of Table 1	DFHS (2) DFS/Int	(1) Int	NonCon	LessInt	Blocked	Random	(1) + (2) Int/Int	NonCon/Int	LessInt/LessInt	Random/Random
ProCon	13,530 (00:04)	**184** (00:00)	9,699 (00:05)	15,736 (00:05)	TO	TO	312 (00:01)	554 (00:01)	**184** (00:01)	TO	310 (00:01)
Rax Extended	**1,288** (00:01)	TO	50,176 (00:10)	TO	TO	TO	TO	TO	TO	TO	OOM
Rax EnvFirst	341,431 (01:08)	OOM	**55,444** (C0:12)	TO	TO	1,541,069 (08:50)	TO	TO	TO	TO	OOM
Readers Writers	3,341 (00:03)	TO	TO	4,087 (00:02)	1,480,875 (03:25)	65,023 (00:09)	1,263 (00:02)	**615** (00:01)	881,410 (02:32)	TO	1,761 (00:01)
Reorder 5 2	**2,891** (00:02)	388,934 (00:39)	–	332,688 (00:34)	297,498 (00:30)	297,683 (00:30)	37,223 (00:07)	–	366,267 (00:39)	297,484 (00:31)	30,725 (00:07)
Sleeping Barber	17 (00:00)	**16** (00:00)	19 (00:00)	18 (00:00)	17 (00:00)	17 (00:00)	88 (00:00)	22 (00:00)	**16** (00:00)	17 (00:00)	65 (00:00)
TwoStage 5 5	**90,321** (00:15)	TO	–	TO	TO	TO	TO	–	TO	TO	4,981,908 (09:47)
Wrong Lock 10 1	459 (00:01)	**56** (00:00)	18,173 (00:05)	104 (00:01)	19,315 (00:06)	17,480 (00:06)	128 (00:00)	86 (00:00)	**56** (00:00)	11,793 (00:05)	120 (00:01)

Table 3 LTL property results

Program	NDFS	DFHS (1)				DFHS (1) + (2)			
		Int	NonCon	LessInt	Random	Int/Int	NonCon/Int	LessInt/LessInt	Random/Random
Account 5	TO	**346,389** (01:25)	4,510,368 (14:59)	TO	TO	536,502 (01:58)	TO	TO	TO
AccountSubtype 2 1	TO	**474,725** (01:33)	4,524,607 (13:56)	TO	–	557,839 (01:52)	TO	TO	–
Airline 5 2	147 (00:01)	12,724 (00:15)	299 (00:03)	147 (00:01)	TO	280 (00:03)	147 (00:03)	**85** (00:00)	TO
AllocateVector	TO	TO	TO	TO	TO	TO	TO	TO	TO
Apps	74,931 (00:19)	20,468 (00:08)	39,881 (00:11)	**1,005** (00:02)	–	400,615 (01:20)	32,664 (00:12)	58,823 (00:18)	–
Deadlock1	268 (00:01)	283 (00:01)	174 (00:01)	268 (00:01)	–	258 (00:00)	**126** (00:00)	188 (00:01)	–
Deadlock2	74 (00:00)	74 (00:00)	72 (00:00)	74 (00:00)	–	34 (00:01)	**32** (00:00)	118 (00:00)	–
DEOS	**87,231** (09:58)	**87,231** (10:03)	**87,231** (10:05)	**87,231** (11:41)	–	**87,231** (10:03)	**87,231** (11:08)	**87,231** (12:11)	–
LinkedList 2 20	TO	TO	**2,640,362** (06:30)	TO	–	TO	TO	TO	–
LoseNotify 3 3 3	**20** (00:00)	55 (00:00)	51 (00:00)	**20** (00:00)	30 (00:00)	48 (00:00)	26 (00:00)	**20** (00:00)	30,120 (00:12)
Piper 1 3 2	TO	**546,163** (01:50)	TO	TO	TO	637,716 (02:01)	TO	TO	TO
ProCon 2 5 6	81,210 (00:21)	603 (00:01)	102,738 (00:25)	TO	TO	642 (00:01)	**196** (00:01)	81,206 (00:19)	TO
Reorder	17,841 (00:07)	**2,276** (00:02)	13,053 (00:07)	17,841 (00:07)	–	5,335 (00:03)	13,027 (00:06)	17,824 (00:07)	–
ReadersWriters 2 2 100	2,561,430 (05:56)	985,973 (02:15)	**6,469** (00:04)	2,477,556 (06:38)	TO	19,244 (00:05)	888,914 (02:42)	2,554,809 (07:38)	TO
TwoStage 1 2	32,725 (00:10)	**7,019** (00:04)	24,854 (00:09)	32,725 (00:11)	–	7,600 (00:05)	25,677 (00:10)	36,507 (00:12)	–
Wronglock 3 2	TO	**165,458** (00:39)	TO	TO	TO	183,377 (00:39)	TO	TO	TO
Daisy	21,862 (00:10)	21,862 (00:10)	2,604 (00:03)	TO	–	**1,522** (00:03)	**1,522** (00:03)	1,635 (00:02)	–

(continued)

Table 3 (continued)

Program	NDFS	DFHS							
		(1)				(1) + (2)			
		Int	NonCon	LessInt	Random	Int/Int	NonCon/Int	LessInt/LessInt	Random/Random
BoundedBuffer 1 1 2 2	2,825,090 (11:52)	**35,845** (00:09)	1,194,051 (05:07)	4,769,245 (09:36)	50,211 (00:12)	40,785 (00:10)	1,649,872 (09:16)	TO	42,366 (00:10)
Clean 1 2 3	OOM	OOM	TO	TO	TO	OOM	OOM	TO	OOM
DiningPhil 3	3,519,876 (06:52)	**211,912** (00:28)	2,272,538 (04:34)	3,116,457 (06:00)	361,666 (00:45)	250,254 (00:34)	2,360,218 (04:47)	3,377,763 (06:36)	358,669 (00:45)
NestedMonitor	<u>62</u> (00:00)	<u>62</u> (00:00)	211 (00:00)	<u>62</u> (00:00)	–	146 (00:00)	146 (00:00)	<u>62</u> (00:00)	–
RaxEnvfirst	TO	TO	TO	TO	TO	TO	TO	TO	OOM
Raxextended	TO	976,131 (01:53)	TO	TO	1,104,188 (01:56)	**721,387** (01:21)	TO	TO	1,432,702 (02:27)
SleepingBarber	TO	**323,565** (00:44)	TO	TO	1,082,807 (02:26)	402,213 (00:54)	TO	TO	954,604 (02:11)
Elevator 4 3 4	OOM	734,636 (01:50)	TO	TO	253,115 (01:05)	**<u>93,903</u>** (00:23)	TO	TO	OOM
AppleOrange 2 2 2	OOM	TO	TO	TO	TO	TO	OOM	TO	OOM
LockThree	1,814,669 (09:20)	**71,715** (01:03)	1,084,153 (07:51)	1,385,777 (07:42)	82,324 (00:52)	105,359 (01:22)	1,225,165 (09:04)	1,515,386 (07:34)	114,803 (01:00)
LockUnlock 3 3 2 2	1,028,749 (05:57)	–	731,249 (04:22)	628,149 (01:58)	**<u>41,791</u>** (00:14)	–	1,171,056 (09:25)	692,361 (02:17)	66,920 (00:21)

Table 4 LTL property results (fairness condition optimization)

Program	DFHS	(1) + (3)				(1) + (2) + (3)			
	(3)	Int	NonCon	LessInt	Random	Int/Int	NonCon/Int	LessInt/LessInt	Random/Random
Daisy	20,742 (00:11)	20,742 (00:12)	29,044 (00:12)	2,284,374 (12:20)	–	418,910 (01:26)	29,453 (00:14)	**1,614** (00:03)	–
BoundedBuffer 1 1 2 2	19,132 (00:13)	1,104 (00:01)	139,976 (00:32)	3,592 (00:03)	–	2,153 (00:02)	**145** (00:01)	3,662 (00:04)	–
Clean 1 2 3	OOM	TO	**655,922** (01:46)	4,009,441 (09:16)	–	TO	TO	5,871,552 (14:35)	–
DiningPhil 3	135 (00:00)	3,998 (00:04)	137 (00:00)	826 (00:01)	–	82 (00:00)	69 (00:00)	**38** (00:00)	8,406 (00:05)
NestedMonitor	46 (00:00)	46 (00:00)	811 (00:01)	46 (00:00)	365 (00:01)	225 (00:00)	755 (00:01)	46 (00:00)	**28** (00:00)
RaxEnvfirst	OOM	TO	TO	TO	TO	**8,700** (00:04)	TO	OOM	TO
Raxextended	4,214 (00:04)	1,582 (00:01)	4,038 (00:02)	307 (00:01)	5,092 (00:04)	1,765 (00:01)	**35** (00:00)	50,271 (00:14)	101,185 (00:22)
SleepingBarber	**28** (00:00)	35 (00:00)	61 (00:00)	**28** (00:00)	460,802 (01:15)	38 (00:00)	**28** (00:00)	**28** (00:00)	842,546 (02:11)
Elevator 4 3 4	TO	203,846 (00:37)	TO	TO	TO	**30,704** (00:11)	TO	TO	OOM
AppleOrange 2 2 2	**408** (00:01)	37,594 (00:14)	1,046 (00:02)	381,397 (01:29)	1,159,729 (07:47)	73,538 (00:23)	2,097 (00:04)	441,587 (01:47)	TO
LockThree	3,575 (00:06)	**651** (00:02)	1,773 (00:04)	2,487 (00:04)	–	15,224 (00:36)	2,313 (00:04)	1,925 (00:04)	20,220 (00:19)
LockUnlock 3 3 2 2	515 (00:02)	–	1,259 (00:03)	5,543 (00:06)	1,500 (00:03)	–	410 (00:02)	**146** (00:01)	31,621 (00:15)

successor orders. The third line indicates which cut-off function and successor order are applied. The best result of each program is underlined.

DHFS and heuristic search are complementary to each other. We do not aim to overcome heuristic search in all cases but show there are cases in which DFHS scores better than current algorithms.

Heuristic search and DFHS are not intended to search all of the state space but to find errors quickly by giving up searching some parts of the state space. Therefore, the heuristic search can end without reporting any errors even when they exist, such as BFS with *Random* heuristic for *Airline* in Table 1.

As shown in Table 1, BFS scored better than DFS in 14 out of 25 programs. On the other hand, DFHS scored the same as or better than DFS in all programs, as shown in Tables 1 and 2. The DFHS algorithm scored better than traditional heuristic search algorithms in 19 out of 25 programs. These results show that DFHS has potential in successfully finding errors in various cases in which traditional heuristic search algorithms fail. For example, DFHS scored more than 20 times better than heuristic search in *Account, AccountSubtype, Airline* and *Piper*.

It is difficult to determine the appropriate policy prior to the execution. However, various policies can be tried in parallel if we have several machines.

6.2 LTL Property

Next, we evaluated DFHS for the LTL property compared with original NDFS. Note that traditional heuristic search is not applicable to NDFS.

The benchmark suite is not designed for LTL verification. Therefore, we modified programs to fit with LTL properties (e.g. removing assert statements, which are obstacles for LTL verification), and set the appropriate LTL formula for each program. We also added more practical programs with LTL formulas.

We applied (1) cut-off function, (2) successor order to all programs, and (3) fairness condition optimization only to the programs that have infinite loops because the fairness condition is meaningful only for such programs.

For each program, we used an LTL formula in the same form as in Sect. 5.2.1. For example, the LTL formula in *Elevator* expresses that an elevator will eventually arrive if a person pushes the call button with a strong fairness condition.

```
([]<>person_enb1->[]<>person_run1)->
([]<>elevator_enb1->[]<>elevator_run1)->
    [](push_down1-><>elevator_arrive1)
```

With the fairness condition optimization of DFHS, we used the following formula and determined if counterexamples conform to the strong fairness condition.

```
    [](push_down1-><>elevator_arrive1)
```

We applied the cut-off functions discussed in Sect. 4.1 and the successor orders discussed in Sect. 4.2. We set the maximum memory size and timeout to be the same as in the safety experiment discussed in Sect. 6.1.

Table 3 lists the results of all programs.[4] In each cell, the number represents the visited states of the synchronized product not the states of the Java virtual machine. Columns labeled with (1) and (2) are the same as in Table 2. Programs from *Account* to *Wronglock* are those that do not have infinite loops. From *Daisy* to *SleepingBarber* are those that have infinite loops. From *Elevator* to *LockUnlock* are the original programs for this study, which have infinite loops and contain liveness bugs. Table 4 lists the results of fairness condition optimization. The label (3) denotes the application of fairness condition optimization and is applied only for the programs that have infinite loops.

With (1) and (2), DFHS was superior to NDFS in 21 out of 28 programs. Additionally, adding (3) improved the results in all cases but *Daisy*. For LTL properties, the size of LTL formula is critical for the entire state space. The application of fairness condition optimization can reduce the state space by omitting the fairness part of the LTL formulas.

7 Conclusion

We proposed a heuristic search algorithm, called DFHS, for efficiently finding errors in software using the software model checker. It is based on DFS but backtracks at the states that have less probability to lead to errors. It uses cut-off functions to determine the necessity to backtrack. It also arranges the order of successors at each state. For LTL search, DFHS reduces the number of states by checking the conformance of the fairness conditions of counterexamples instead of specifying the fairness conditions in LTL formulas.

We enhanced JPF, a software model checker for Java byte code, to implement DFHS and NDFS. We evaluated DFHS with programs and compared it with current search algorithms. The experimental results show that DFHS scored better than the current algorithms in many programs.

For future work, we plan to apply DFHS, which uses many parameters, cut-off functions and thresholds, to larger systems. We will enhance our approach to dispatch search tasks to several model checkers using DFHS with different parameters that concurrently search for counterexamples.

---

[4]We omitted *AlarmClock* from the table, which throws an uncaught exception.

References

1. Barnat, J., Brim, L., Chaloupka, J.: Parallel breadth-first search LTL model-checking. In: Proceedings of 18th IEEE International Conference on Automated Software Engineering, pp. 106–115. IEEE Computer Society (2003)
2. Concurrency tool comparison site. https://facwiki.cs.byu.edu/vv-lab/index.php/Concurrency_Tool_Comparison
3. Courcoubetis, C., Vardi, M.Y., Wolper, P., Yannakakis, M.: Memory efficient algorithms for the verification of temporal properties. In: Proceedings of 2nd International Workshop on Computer Aided Verification. LNCS, vol. 531, pp. 233–242. Springer (1990)
4. Couvreur, J.M.: On-the-fly verification of linear temporal logic. In: Proceedings of World Congress on Formal Methods. LNCS, vol. 1708, pp. 253–271. Springer (1999)
5. Couvreur, J., Duret-Lutz, A., Poitrenaud, D.: On-the-fly emptiness checks for generalized Büchi automata. In: Proceedings of 12th International SPIN Workshop on Model Checking Software. LNCS, vol. 3639, pp. 169–184. Springer (2005)
6. Cuong, N.A., Cheng, K.S.: Towards Automation of LTL Verification for Java Pathfnder. National University of Singapore (2008)
7. Duret-Lutz, A., Poitrenaud, D., Couvreur, J.M.: On-the-fly emptiness check of transition-based Streett automata. In: Proceedings of 7th International Symposium on Automated Technology for Verification and Analysis. LNCS, vol. 5799, pp. 213–227. Springer (2009)
8. Edelkamp, S., Lluch-Lafuente, A., Leue, S.: Trail-directed model checking. Electron. Notes Theor. Comput. Sci. **55**(3), 343–356 (2001)
9. Edelkamp, S., Schuppan, V., Bosnacki, D., Wijs, A., Fehnker, A., Aljazzar, H.: Survey on directed model checking. In: Proceedings of 5th International Workshop on Model Checking and Artificial Intelligence. LNCS, vol. 5348, pp. 65–89. Springer (2008)
10. Geldenhuys, J., Valmari, A.: More efficient on-the-fly LTL verification with Tarjan's algorithm. Theor. Comput. Sci. **345**(1), 60–82 (2005)
11. Groce, A., Visser, W.: Heuristics for model checking Java programs. Int. J. Softw. Tools Technol. Transf. **6**(4), 260–276 (2004)
12. Holzmann, G.J.: The SPIN Model Checker—Primer and Reference Manual. Addison-Wesley, Boston (2004)
13. Lombardi, M.: https://bitbucket.org/michelelombardi/jpf-ltl
14. McMillan, K.L.: Symbolic Model Checking. Kluwer, Boston (1993)
15. Musuvathi, M., Qadeer, S.: Iterative context bounding for systematic testing of multithreaded programs. In: SIGPLAN 2007 Conference on Programming Language Design and Implementation, pp. 446–455. ACM (2007)
16. Parízek, P., Lhoták, O.: Randomized backtracking in state space traversal. In: Proceedings of 18th International SPIN Workshop. LNCS, vol. 6823, pp. 75–89. Springer (2011)
17. Renault, E., Duret-Lutz, A., Kordon, F., Poitrenaud, D.: Three SCC-based emptiness checks for generalized Büchi automata. In: Proceedings of 19th International Conference on Logic for Programming, Artificial Intelligence, and Reasoning. LNCS, vol. 8312, pp. 668–682. Springer (2013)
18. Rungta, N., Mercer, E.G.: A meta heuristic for effectively detecting concurrency errors. In: Proceedings of 4th International Haifa Verification Conference. LNCS, vol. 5394, pp. 23–37. Springer (2008)
19. Rungta, N., Mercer, E.G.: Clash of the Titans: tools and techniques for hunting bugs in concurrent programs. In: Proceedings of 7th Workshop on Parallel and Distributed Systems: Testing, Analysis, and Debugging. ACM (2009)
20. Russell, S.J., Norvig, P.: Artificial Intelligence—A Modern Approach, 3rd edn. Pearson Education, Upper Saddle River (2010)

21. Sun, J., Liu, Y., Dong, J.S., Pang, J.: PAT: towards flexible verification under fairness. In: Proceedings of 21st International Conference on Computer Aided Verification. LNCS, vol. 5643, pp. 709–714. Springer (2009)
22. Vardi, M.Y., Wolper, P.: Automata-theoretic techniques for modal logics of programs. J. Comput. Syst. Sci. **32**(2), 183–221 (1986)
23. Visser, W., Havelund, K., Brat, G.P., Park, S., Lerda, F.: Model checking programs. Autom. Softw. Eng. **10**(2), 203–232 (2003)

A Novel Architecture for Learner's Profiles Interoperability

Leila Ghorbel, Corinne Amel Zayani and Ikram Amous

Abstract Generally, many adaptive systems are developed and used in various fields. The effort to build the user's profile is repeated from one system to another due to the lack of interoperability and synchronization. Therefore, to provide an effective interoperability is a complex challenge due to the evolution of the user's profiles and its heterogeneity. The user's profiles evolution is not taken into account in the interoperable system. In our work, we are interested in the educational field. In this context, we propose a novel interoperable architecture allowing the exchange of the learner's profile information between different adaptive educational cross-systems to provide an access corresponding to the learners' needs. This architecture is automatically adapted to the learner's profiles that evolve over time and are syntactically, semantically and structurally heterogeneous. An experimental study shows the effectiveness of our architecture.

1 Introduction

The adaptive systems represent the user's profile in different formats (different representation standard, different syntactic and semantic representation, etc.). Each system can have incomplete or partial user's information: the profile can be empty or contain very little information, so no user's adaptation can be realized. This problem is known in the literature as the cold-start problem. Hence, we need to exchange (share) the user's profiles (or parts) between different systems to enhance and integrate the user's

L. Ghorbel (✉) · C.A. Zayani · I. Amous
MIRACL-ISIMS Sfax University, Tunis Road Km 10, 3021 Sfax, Tunisia
e-mail: leila.ghorbel@gmail.com

C.A. Zayani
e-mail: corinne.zayani@isecs.rnu.tn

I. Amous
e-mail: ikram.amous@isecs.rnu.tn

© Springer International Publishing Switzerland 2016
R. Lee (ed.), *Computer and Information Science 2015*,
Studies in Computational Intelligence 614, DOI 10.1007/978-3-319-23467-0_7

knowledge (profile enrichment). This is a kind of cooperation between adaptive systems to provide more results adapted to the user's expectations. This cooperation requires having interoperable user's profiles for different systems.

The exchange of the user's profiles data between different systems is the main phase to get interoperable systems. However, other phases must be really highlighted before and after this phase (system discovery, user's identification and evaluation of the exchanged data). Several Works in different fields have been proposed in literature to provide solutions to the problem of the user's profiles interoperability [8, 12, 13]. These works mainly provide structure, semantics and a common model of the user's profiles for all systems in order to simplify the profile exchange. However, to our knowledge, these common models do not resolve the problem of data exchange between profiles evolving over time. In our work, we are interested in the user's profiles interoperability in the educational field. Therefore, the purpose of this paper is to propose a novel architecture with a common profile called Global Profile allowing the exchange of learner's profiles between different adaptive educational cross-systems to provide an access corresponding to the learner's needs. These profiles evolve over time and are syntactically, semantically and structurally heterogeneous.

In this paper, we first present the general concepts of interoperability. Then, in Sect. 3 we present a state of the art on the works that specifically address the interoperability of the user's profiles in different fields. In Sect. 4, we present our contribution in the proposal of an interoperable architecture for the exchange of data between user profiles that evolve over time in adaptive educational cross-systems. In this section, we focus on the main process of this architecture. In Sect. 5, we describe the results of the evaluation of our architecture. We end up with a conclusion and an overview of the ongoing works that we try to achieve.

2 General Concepts

As mentioned in the introduction, we are interested in the problem of the user's profiles Interoperability. One of the definitions of interoperability is the one given by Wegner who defines interoperability as the ability to cooperate and exchange data despite the differences between the languages, interfaces, and execution platforms [1, 4]. In order to overcome such differences, we must resolve them in the structure, syntax and language. In literature, three types of interoperability are defined [5]: structural, syntactic and semantic. Structural interoperability concerns the possibility of reducing the differences between the systems at the access level (communication protocols, standardized interfaces for accessing data …). The syntactic interoperability is related to the ability of different systems to interpret the syntax of the data in the same way. The semantic interoperability is related to the ability of different systems to interpret the semantics of the data in the same way.

This definition of interoperability may be revised to get the definition for the interoperability of the user's profiles. Several authors [10, 14] defined the user's profiles interoperability by means of: (1) accelerating the initialization of the user's profiles in the case of the cold-start problem, (2) acquiring relevant user data, and (3) exchanging the user's profiles.

The user's profile interoperability takes place in four phases [5]. The first phase is to discover the systems that store data about a specific user. In the second phase, the systems must agree on the identification of the user then, the data should be exchanged at the third phase and evaluated qualitatively and quantitatively at the fourth phase.

When dealing with an interoperable system, several details must be highlighted among which we can mention: the main task of interoperability and its architecture. Three main interoperability tasks (categories) have been distinguished in literature [5]. There are systems that aim to facilitate the exchange on request of the user's profiles. This is used when the systems have incomplete information about the user. Other systems provide a service of the user's profile adaptation or modeling at real-time when receiving data from other systems. This is useful when the systems that require the user's profile do not have a user's profile or adaptation functions. These systems do not share the user's profile data with other systems. On the contrary, the latter systems share user's profiles. They collect data about the user and integrate them to form a richer user's profile to make it available for the other systems.

There are three architectures for the user's profiles interoperability: centralized, decentralized and mixed [5]. In these architectures, we find the following main components: systems, user's profiles and user's profile storage units. In the centralized architecture, all the user data are stored in a central storage unit. The user's profile is unique and centralized: Generic User Profile PUG. The advantage of this architecture is that the data of the user's profile are available for multiple systems. However, the PUG is restrictive and cannot contain all of the user's data. In the decentralized architecture, each system is occupied by the management of its local user's profile storage and communicates with other systems to collect the required data. In this architecture, each system is independent, but the connection between the systems is one to one. In mixed architecture, each system has a local user's profile storage unit (decentralized approach) which refers to a central user's profile (centralized approach). This approach provides more flexibility and applicability to the user's profiles, but many conflicts or redundancies in the collected data may occur.

3 State of the Art

Several works have been proposed in literature to provide solutions to the problem of the user's profiles interoperability. We are specifically interested in the work proposed for the problem of the third phase which is the user's profile data exchange between different systems. These systems represent the user's profiles differently at the syntax and the structure level. In order to exchange data between these different profiles, some dimensions that we have learned from the analyzed work must be respected. These dimensions, which are shown in Table 1 (see Sect. 3.3), are the task of interoperable system, the architecture, the representation of the exchanged data, the languages and communication protocols and the integration of the exchanged data and the type of the exchanged data. Our work focuses mainly on the dimension of the representation and the integration of the exchanged data.

3.1 Representation of the Exchanged Data

This dimension describes how the data are represented at the exchange phase. In literature, three representations are distinguished: (i) a standardized user's profile representation (common representation) using ontology or unified profiles [1, 12] (ii) a second based on mediation (translation) of different profile data of a specific user [8] and (iii) a third based on the two first representation [2, 3, 6, 13].

The first representation is based on the definition of standard ontology or unified profiles that can be used by multiple systems. The Work in [8] provides a common basis for the exchange of the learner's profiles between several educational systems. This work is based on standards for the learner's profiles modeling. Thus, an ontology learning field is used to share the learner's profiles. The work in [12] provides a common representation (XML) of two learner profile standards (PAPI and IMS) through which the data are exchanged. PersonisAD [1] is an approach for building a user's profiles by distributed applications using a common semantic profile.

This representation is an incomplete solution because of the diversity of systems (educational, recommendations …), variety of the stored user's profile data (interests, preferences, navigation historic, evaluation …) and the large quantity of differences in syntax, structure and semantic of the latter ones. In fact, the emergence of a new system requires the reconstruction of a new ontology or a new unified profile. For this reason, the second representation appeared to be a possible solution to these problems.

The solution is to use mediation techniques [16] to develop a mapping between the different representations of a user's profiles using suitable mapping rules. The data exchange is made after the achievement of the semantic agreement. Each system in this category has its own user's profile and creates a central profile model containing the most frequently used data that can be shared.

To exchange data, each system must do the mapping with the central models of other systems. This requires a mediation process (or translation) which is carried out through a component called "mediator". For example, the authors in [4] use techniques of automatic and semi-automatic mapping to convert data between the user's profiles.

This category has the advantage that each system may adopt its own representation of the user's profiles in terms of language and structure. However, each peer system needs to create the mapping in both directions. Therefore, the mediator must implement multiple mapping rules for different user's profile models. In addition, during the introduction of a new user's profile model representation, the mediator must develop new mapping rules from this representation to other representations and vice versa.

The third representation appeared to overcome the disadvantages of the first two. The representation of the exchanged data is based on a common user's model and the translation of data between the common and the other system models. Among the works that adopted this category, we cite [2, 6, 13]. For example, the data exchange in the FUSE [13] approach which is conducted through a canonical model with a translation process based on mappings.

3.2 Integration of the Exchanged Data

This dimension describes how the exchanged data can be integrated and how to deal with generated conflicts. In literature, we distinguish two data integration approaches.

In the first, the data collected from different systems are not merged (without fusion) in each existing user's profile, but they are used only when needed. The majority of works in literature adopt this approach. They consider that interoperable systems are responsible for the integration of the exchanged data as needed. In [2], the data are stored in different systems and transferred, on request, to the mediator. Then, they are converted into the requested format but not stored in the mediator. Thus, the integration is done on the fly when needed.

In the second approach, the collected data are merged (fusion) into the existing user's profiles. This approach requires data integration and conflict resolution operations. The conflicts can occur in the data or the data values collected from the users profiles of different systems. There is a research that supports the integration of the exchanged data with fusion and conflict resolution [4, 13]. The authors in [4] tried to merge the exchanged data and resolve conflicts in data values from different systems. The conflict resolution is done through the measure of credibility of the exchanged data and of the system suppliers of these data. In the FUSE [13] approach, fusion is the result of data mappings, and the technologies used for data exchange. In fact, the mapping and the conflict detection are performed manually by an administrator.

3.3 Synthesis

To summarize, we present some user's profiles interoperability works in a table describing the main dimensions previously seen (see Table 1). This table shows that most of the works solve the problem of interoperability through the exchange of user's profiles in a decentralized manner [4, 6, 8]. This may raise a few problems among which we can mention: the possibility of producing data redundancies, the difficulty of achieving syntactic and semantic interoperability and the definition of parts of the user's profile to be shared. Other works tried to solve the problem of interoperability through the user's modeling service in a centralized manner and through a common user's profile model [1, 3]. In this case, the development of full domain ontology or the modeling of the user's profiles in all possible contexts is an unrealizable solution. This is the cause of the large number of syntactic and structural

Table 1 User's profiles interoperability works

Approach	Interoperability task	Architecture	Protocols and languages	Exchanged data representation	Exchanged data Integration	Exchanged data
Ref. [8]	User profile exchange	Decentralized	Web service RDF	Common model	Fusion of the collected data	Competence, preferences, learning style, certificates, …
Ref. [12]		Mixed	Web service RDF	Common model	Fusion of the collected data	Preferences, personal data, …
Ref. [13]		Mixed	REST/XML	Common model and translation	Fusion of the collected data	Social data, personal data, learning style …
Ref. [3]	User profile Modeling service	Centralized	HTTP/XML	Common model and translation	No	Personal data, interests, interactions historic, …
Ref. [4]	User profile exchange	Decentralized	HTTP/RDF, RDFS	Translation	Fusion of the collected data	Personal data, interests, preferences, …
Ref. [6]		Decentralized	SOAP/RDF, RDFS	Common model and translation	No	Interests
Ref. [2]		Mixed	Web service	Common model and translation	No	Social data, personal data, …
Ref. [1]	User profile modeling service	Centralized	HTTP/JSON	Common model	No	Interests, preferences, …

differences between the models of each user's profile. To solve these problems, some works proposed to use mixed architecture.

We can see from the analyzed works that the most widely used protocols are HTTP, SOAP or REST and the most used language representations are XML or RDF. These works tend to solve the problem of semantic interoperability because the data exchange is done between the user's profiles represented by the same language. However, the user's profile is not always represented from a system to another by the same language (syntax). As a result, the problem of syntactic interoperability must be solved.

We also note that most of the works proposed a hybrid solution for the representation of the exchanged data with fusion [13] or without fusion [2, 3, 6]. This solution is based on the advantages of the common user's profile model representation and mediation (translation). To mediate the data, it is necessary to make the mappings between the different user's profiles. This is done either by the use of the generic mapping tools like Altova-MapForce[1] or manually. In both cases, it is difficult to create a suitable mapping when dealing with a large amount of semantic and syntactic heterogeneity.

Despite the solutions offered in the different approaches, further efforts must be made to resolve the problems already mentioned and other problems that are not yet resolved. In particular, it is necessary to consider that the user's profile evolves with the exchange and integration of the data. Therefore, the part or parts of the profile to be shared, used and modified by other systems should be known. In fact, over time, and after numerous exchange and integration operations, the user's profile can be overloaded and the system cannot distinguish relevant data to take into account to better respond to the needs of the user and exchange with other systems. Consequently, each system may be faced with a cognitive profile overload. To solve the problem mentioned above, the profile should always contain the most relevant data to be shared or exchanged. This problem should also be solved in the generic user's profile through which the systems exchange data.

Given the limitations of the work discussed, we propose an interoperable architecture allowing the exchange of the user's profile information between educational cross-systems in order to provide a better uniform access to the user's profiles that evolve over time and are syntactically, semantically and structurally heterogeneous to get access to the corresponding user's needs.

4 Proposed Architecture

Our architecture aims to resolve the problems related to the interoperability process (see Sect. 3). Mainly, this resolve focuses on the exchange data phase between adaptive educational cross-systems. In the educational adaptive system, the learner's profile constitutes a key element to improve the learner's adaptation results. Figure 1 shows the proposed architecture.

---

[1] www.altova.com/MapForce.

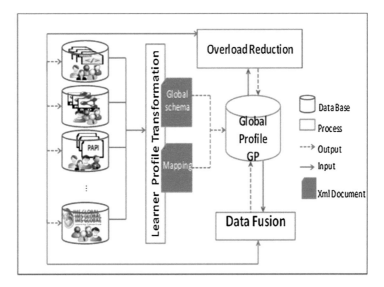

Fig. 1 The proposed architecture

This mixed architecture takes place in three main processes. As mentioned previously (see Sect. 2), in the mixed architecture, each system has a local user's profile storage unit which refers to a central user's profile over mapping techniques. Therefore, we find these two classical processes: the Learner Profile Transformation and the Data Fusion. These processes are related to the Global Profile GP.

The first process is for transforming the local learner's profiles which are represented by different standards (XML, RDF), structures and semantics to a global one Global Profile GP by creating automatic mapping rules to resolve such differences and a global schema. The mapping rules and the global schema are represented in xml.

In order to exchange data, each system is expected to map to the GP based on the global schema and the mapping rules. Then, the exchanged data may or may not be merged in GP. In our work, we merge the exchanged data in GP based on the Data Fusion process. In the exchanged data coming from different systems, possible conflicts may occur. For this reason, these conflicts should be resolved with the Data Fusion process.

With the first user's interaction, each system can have incomplete or partial user's data: the leaner's profiles can be empty or contain very little information. Therefore, these systems need to exchange data stocked in GP to solve the problem of the cold start problem. For this purpose, the exchanged data should be merged in the local learner's profiles.

Thus, the recurrent user-system interaction means several data fusion operations in GP and also in the local profiles. As a result, the GP and the local profiles evolve over time and become overloaded with pertinent and non pertinent data. In addition, to know which pertinent data of local user's profiles should be merged in the GP and which part of GP should be considered to better respond to the user's needs (expressed by request, historic navigation), we propose to add a new process called Overload Reduction. Some methods have conducted to resolve the problem of the user's profile overload but not in interoperable systems. These methods can be grouped in two categories: implicit and explicit according to the user's intervention. Explicit methods require the user's intervention to remove the non pertinent data from his profile [11], while the implicit ones are automatic machine-based learning techniques [7, 9].

This process is based on the method proposed in our previous work [17] which is based on the semi-supervised learning technique and specifically on Co-Training algorithm to detect and remove non pertinent data. This method is automatically adapted to the content of any profile. An experimental study by qualitative and comparative evaluations shows that this method can detect and remove non pertinent profile data effectively [18].

5 Experimentation

In order to prove the utility of our architecture, we select two distributed educational system: (1) a learning management system called Moodle and (2) a learning assessment system called Position Platform. The learners are members of both systems at the same time. In Moodle, the learners can learn courses, do activities, receive marks about these activities, take exams, etc. In the Position Platform, the learners can take exams in the form of multiple choice questions and get marks. The learner's profiles are represented in different structure, syntax and semantics. The data exchange between the two learner's profiles is benefic for many reasons. The Sara's case can demonstrate some reasons.

Sara studies in the Virtual University of Tunis. She wants to get a certificate in Information technologies and Internet (C2I). The courses of this module belong to 5 domains and each domain includes 4 competencies (sub-domains) where each one includes several themes. Sara wants take the certificate exam. When she accesses her count on Moodle for revising, she faces several links related to the whole domains (competencies, activities, etc.) in which some links are not useful. This is due to the overload of the Sara's profiles with pertinent and non pertinent data describing the whole learning experience.

Sara needs to be oriented with the best links to accomplish her revision: links to domains (or sub_domains) that Sara doesn't have the score average in the related activities and the related passed exams.

The score exchange of Sara in such domain (competency) in the Position Platform deals with such difficulties because the learner's profiles are different. The learner's profile in Moodle is represented by the IMS standard and respects the XML syntax. In the Position Platform, it is represented by the PAPI standard and respect RDF syntax. Therefore, the learner's profiles evolve after each learner-system interaction. For this reason, the proposed Global Profile and the Overload Reduction process

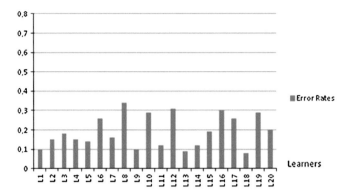

Fig. 2 Evaluation of the error links sequencing

must be setting up to resolve these problems. The global profile should contain the overall pertinent parts of each local profile to provide an access corresponding to the learners' needs (links to revise the certificate).

The evaluation is performed on 40 learners based on two values: the error rate and the success percentage.

The first 20 learners revise for the C2I certificate in which our architecture is not setting up. The second 20 revise for the C2I certificate based on links recommended by our architecture. The evaluation error rate is based on the first 20 learners (see Fig. 2).

$$ERROR - RATE = \frac{\text{NB-ERROR}}{\text{NB-RECOM-LINK}} \tag{1}$$

To calculate the error rate (see Eq. 1), we compare the visited link sequence when a learner wants to revise for the certificate and the recommended link sequence which is based on our architecture. The error rate (ERROR-RATE) is the number of errors in link sequence (NB-ERROR) divided by the total number of the recommended links (NB-RECOM-LINK).

As we can see in Fig. 2, the most of the learner's error rate values are low. Actually, they are comprised between 0,1 and 0,35 and the average error rate is 0,19. This average error explains that 89 % of the recommended links are pertinent and correspond to the learners' needs (link sequence).

After the revision, the learners took the certification exam. The results show a clear improvement in the success percentage: 65 % of the first 20 learners and 90 % of the second 20 ones. The evaluation values confirm the effectiveness of our architecture.

6 Conclusion

In this paper, we presented our architecture that solves the problems associated with the exchange of the learner's profile data in adaptive educational cross-systems. This architecture offers a unified and transparent access to the different learner's profiles that evolve over time over the Global Profile. This architecture was evaluated on 40 learners' profiles and showed good results. In our future works we will focus on the application of our architecture in educational cross-systems and social networks in order to exchange interests and preferences to improve the effectiveness of results.

References

1. Assad, M., Carmichael, D.J., Kay, J., Kummerfeld, B.: PersonisAD: distributed, active, scrutable model framework for context-aware services. Pervasive Computing. Lecture Notes in Computer Science, pp. 55–72. Springer, New York (2007)
2. Berkovsky, S., Kuflik, T., Ricci, F.: Mediation of user models for enhanced personalization in recommender systems. User Model. User-Adapt. Interact. 18(3), 245–286 (2008)
3. Brusilovsky, P., Milln, E.: User models for adaptive hypermedia and adaptive educational systems. Adapt. Web, Methods Strat. Web Personal. 4321, 3–53 (2007)
4. Carmagnola, F., Dimitrova, V.: An evidence-based approach to handle semantic heterogeneity in interoperable distributed user models. In: Proceedings of the 5th International Conference on Adaptive Hypermedia and Adaptive Web-Based Systems Lecture Notes in Computer Science, pp. 73–82 (2008)
5. Carmagnola, F., Cena, F., Gena, C.: User model interoperability: a survey. User Model. User-Adapt. Interact. 21(3), 285–331 (2011)
6. Cena, F.: Integrating web service and semantic dialogue model for user models interoperability on the web. J. Intell. Inf. Syst. 36(2), 131–166 (2011)
7. Chunyan, L.: User profile for personalized web search. Fuzzy Systems and Knowledge Discovery FSKD, 1847–1850 (2011)
8. Dolog, P., Schfer, M.: A framework for browsing, manipulating and maintaining interoperable learner profiles. User Model. 3538, 397–401 (2005)
9. Eypharabid, V., Amandi, A.: Ontology-based user profile learning. Appl. Intell. 36, 857–869 (2012)
10. Kobsa, A.: Generic user modeling systems. Adapt. Web, Methods Strat. Web Personal. 49–63 (2001)
11. Mandeep, P., Rachid, A., Michael, O., Anne, J.: Explicit user profiles in web search personalisation. In: Proceedings of International Conference on Computer Supported Cooperative Work in Design, pp. 416–421 (2011)
12. Musa, D.L., Oliveira, JPMde.: Sharing learner information through a web services-based learning architecture. J. Web Eng., pp. 263–278 (2005)
13. Walsh, E., O'Connor, A., Wade, V.: The FUSE domain-aware approach to user model interoperability: a comparative study. Inf. Reuse Integr. 554–561 (2013)
14. Wang, Y., Cena, F., Carmagnola, F., Cortassa, O., Gena, C., Stash, N., Aroyo, L.: Rss-based interoperability for user adaptive systems. Adapt. Hypermedia Adapt. Web-Based Syst. 353–356 (2008)
15. Wegner, P.: Interoperability. Comput. Surv. 28, 285–287 (1996)
16. Wiederhold, G.: Mediators in the architecture of future information systems. IEEE Comput. Mag., pp. 38–49 (1996)

17. Zghal, R., Ghorbel, L., Zayani, C.A., Amous, I.: An adaptive method for user Profile learning. In: ADBIS, Advances in Databases and Information Systems, pp. 126–134 (2013)
18. Zghal, R., Ghorbel, L., Zayani, C.A., Amous, I.: Pertinent user profile based on adaptive semi-supervised learning. In: KES, Knowledge Based and Intelligent Information and Engineering Systems, pp. 313–320 (2013)

CORE: Continuous Monitoring of Reverse k Nearest Neighbors on Moving Objects in Road Networks

Muhammad Attique, Hyung-Ju Cho and Tae-Sun Chung

Abstract A reverse nearest neighbor (RNN) query retrieves all the data points that have q as one of their closest point. In this paper, we study the problem of a continuous reverse nearest neighbor queries where both query object q and data objects are moving. We present a new safe exit based algorithm for efficiently computing safe exit points of query and data objects for continuous reverse nearest neighbor queries called CORE. Within the safe region, query result remains unchanged and a request for recomputation of query does not have to be made to the server. This significantly improves the performance of algorithm because the expensive recomputation is not required as long as the query and data objects are within their respective safe exit points.

Keywords Continuous monitoring · Reverse nearest neighbor query · Road network · Safe exit algorithm

1 Introduction

The rapid development of GPS-based devices and location based environment has been growing in the past decades. These systems enabled the existence of real world applications such as the retail services, mixed reality games, army strategy planning and enhanced 911 services. The continuous movement of data objects demands for the new query processing techniques to cater the frequent location updates. While a plethora of work have been devoted to moving query processing [4, 5, 12] they

M. Attique · T.-S. Chung (✉)
Department of Computer Engineering, Ajou University, Suwon 443-749, South Korea
e-mail: tschung@ajou.ac.kr

M. Attique
e-mail: attique@ajou.ac.kr

H.-J. Cho
Department of Software, Kyungpook National University, Gajang-dong,
Sangju-si, Gyeongsangbuk-do 742-711, South Korea
e-mail: hyungju@knu.ac.kr

© Springer International Publishing Switzerland 2016
R. Lee (ed.), *Computer and Information Science 2015*,
Studies in Computational Intelligence 614, DOI 10.1007/978-3-319-23467-0_8

all focus on range queries and nearest neighbor (NN) queries, there is still a lack of research in addressing continuous reverse nearest neighbor (RNN) queries. In addition, majority of research has been conducted on the Euclidean space, not on the road network.

Consider a query object q and a set of data objects O (e.g., restaurants and gas stations). We use $sdist(o, q)$ and $sdist(o, o')$ to represent the shortest distance from object o to query q and another data object o', respectively. A reverse k nearest neighbor (RkNN) query retrieves all the data objects that are closer to q than to any other data object o' i.e., $sdist(o, q) < sdist(o, o')$.

RNN queries are generally categorized into two types: monochromatic reverse kNN (MRkNN) queries and bichromatic reverse kNN (BRkNN) queries [6, 7]. In the monochromatic RNN, all moving data and query objects are of the same type. Applications of the continuous monochromatic RNN include mixed reality games in which the goal of each player is to shoot the player nearest to him. Each player needs to continuously monitor his own reverse nearest neighbor to avoid being shot by other players. In the bichromatic RNN, query objects and data objects belong to two different types of objects. Applications of the continuous bichromatic RNN include army strategy planning where a medical unit might issue a bichromatic RNN query to find closest wounded soldiers.

In general, the main challenge for continuous monitoring algorithms is how to maintain the freshness of query results when the query and data objects move freely and arbitrarily. A simple approach is to increase the frequency of updates where query q periodically sends requests to re-evaluate the query results. However, this approach still does not guarantee that results are fresh because the query results may still become stale in between each call to the server. In addition, excessive computational burden may be imposed on the server side with a high communication frequency imposed on communication cost.

To address the aforementioned issue, we present a safe exit based approach for continuous monitoring of reverse k nearest queries in a road network where both query objects and data objects are moving arbitrarily in a road network. Our algorithm computes safe exit points for both query and data objects. The query result remains unchanged as long as query and data objects lies within their respective safe exit points. The safe exit technique avoids the back and forth communication between client and server resulting in cutting down the communication and computation cost.

Specifically, our contributions can be summarized as:

- We present a framework for continuous monitoring of RkNN queries where both query and data objects are moving in road networks.
- We present pruning rules that optimize the computation of safe exit points by minimizing the size of unpruned network and number of objects.
- An experimental study confirms that our approach outperforms a traditional approach in terms of both communication and computation costs.

Related work is first reviewed in Sect. 2, followed by introduction of terminology definitions and describes the problem. Section 4 elaborates on our proposed safe exit algorithm (CORE) for computing the safe exit points of moving RkNN queries in road networks. Section 5 presents a performance analysis conducted of the proposed technique. Section 6 concludes this paper.

2 Related Work

An RNN query for moving objects searches for those objects that take query object q as their nearest neighbor. The processing of RNN queries has become one of the recent emerging areas of research. Korn et al. [7] were the first to introduce the concept of RNN queries. They used the pre-computing technique to search for RNNs. One of the important categories of RNN query algorithms query is continuous RNN query algorithms, which can incrementally give the RNN results. The existing continuous query processing methods focuses on defining the monitoring region and updates the query results based on the moving object's location. Benetis et al. [1] were the first to study continuous RNN monitoring, but their proposed scheme assumes that the velocity of objects are known.

In recent years, reverse neighbor query processing in road network has received significant attention by the spatial database systems research community. Yiu et al. [13] first addressed the issue of RNN in road networks and proposed an algorithm for both monochromatic and bichromatic RkNN queries. Safar et al. [10] presented a framework for RNN queries based on network voronoi diagrams (NVDs) to efficiently process RNN queries in road networks. However, their scheme is not suitable for continuous RNN queries due to the fact that an NVD changes whenever a dataset changes its location, resulting in high computation costs.

Sun et al. [11] studied a continuous monitoring of bichromatic RNN queries. They associated a multiway tree with each query to define the monitoring region, and only the updates in the monitoring region affect the results. However, this method is limited to bichromatic queries and also does not cater for $k > 1$. Moreover, their proposed scheme assumes that the query objects are static. Li et al. [8] proposed a novel algorithm for continuous monitoring of RkNNs based on a dual layer multiway tree (DLM tree) in which they introduced several lemmas to reduce the monitoring region and filter the candidate objects. Their continuous monitoring of RkNN method comprises two phases: the initial result generating phase and the incremental maintenance phase. Cheema et al. [3] proposed a safe region approach for the monitoring of continuous RkNN queries in Euclidean and road networks. However, to provide the safe region (which may consist of complex road segments) more network bandwidth is consumed compared to simply providing a set of safe exit points representing the boundary of the safe region.

3 Preliminaries

Section 3.1 defines the terms and notations used in this paper, while Sect. 3.2 provides a problems description with the help of example.

3.1 Definition of Terms and Notations

Road Network: A road network is represented by a weighted undirected graph $G = (N, E, W)$ where N shows the set of nodes $N = \{n_1, n_2, \ldots, n_{|N|}\}$, E is a set of edges that connects two distinct nodes $E = \{e_1, e_2, \ldots, e_{|E|}\}$, and $W(e)$ denotes the weight of an edge e. An edge between two nodes is denoted by $e(n_s, n_e)$, where n_s and n_e are referred as boundary nodes.

   Segment: Segment $s_{(p_1, p_2)}$ is the part of an edge between two points, p_1 and p_2, on the edge. An edge consists of one or more segments.

   Figure 1 shows an example of an undirected road network with six nodes, n_1 to n_6. Several edges and segments are shown with their respective weights. For example, the edge $e(n_1, n_2)$ consists of segments $s_{(n_1, o_1)}$ and $s_{(o_1, n_2)}$ having weights 3 and 2, respectively. There are six data objects in this example $\{o_1, o_2, \ldots, o_6\}$ and single query object q. Query q and data objects are shown as triangle and rectangles, respectively. Given two points p_1 and p_2 the shortest path distance $sdist(p_1, p_2)$ is the minimum distance between p_1 and p_2. In Fig. 1, shortest path from q to o_3 is $q \rightarrow n_3 \rightarrow n_5 \rightarrow o_3$ and $sdist(q, o_3) = 10$.

3.2 Problem Description

In this paper, we primarily address the problem of continuous monitoring of RkNN queries on moving queries and data objects in road networks. To provide a clear

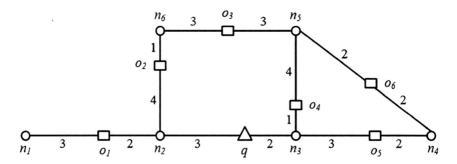

Fig. 1 Example of a road network

Table 1 Summary of notations used in this paper

Notation	Definition
G = (N, E, W)	Graph model of a road network
dist(p_s, p_e)	Length of the shortest path from p_s to p_e, where p_s and p_e represents start and end points, respectively.
n_i	A node in the road network
$e(n_s, n_e)$	An edge in the edge set E, where n_s and n_e are start and end points of edge represented as boundary node (β), such that $\beta \in \{n_s, n_e\}$
W(e)	Weight of the edge $e(n_s, n_e)$
q	A query point in the road network
k	The number of requested RNNs
O	The set of objects $O = \{o_1, o_2, \ldots, o_n\}$
O^+	Set of answer objects $O^+ = \{o_1^+, o_2^+, \ldots, o_n^+\}$
O^-	Set of non-answer object $O^- = \{o_1^-, o_2^-, \ldots, o_n^-\}$
IO^+	Influence region of answer objects
IO^-	Influence region of non-answer objects
p_{anchor}	An anchor point that corresponds to the start point of expansion
$o_{farthest}^+$	The farthest answer object to a point $p \in G$
$o_{nearest}^-$	The nearest non-answer object to a point $p \in G$
p_{se}	A safe exit point where the safe and non-safe region q or o intersects.
R_p	The set of RkNNs at a point p

explanation, we use the road network example shown in Fig. 1, in which there are six objects, o_1 to o_6, and a query q in a road network (Table 1).

For sake of explanation, we are considering the monochromatic RNN queries where both data objects and query objects belong to the same data type. However, our method can extend to monitor continuous bichromatic RkNNs queries too.

Let us assume that a moving query requests one RNN ($k = 1$) at certain point p_1. In order to get one RNN, we traverse the road network from active edge which contains point q. For each data object $o \in O$ encountered, issue a verification query verify (o, k, q) that check whether it is RNN or not. If there exists another object o' such that $dist(o, o') < dist(o, q)$ then o is not a RNN. Otherwise, o is inserted into the RNN result set denoted as R_p. The expansion in each path stops once k objects have been found in that direction. As for getting RNN at point P_2, the simple approach is to repeat the procedure executed at p_1. However, this recomputation whenever query q or any data object changes its location significantly degrades the performance of the algorithm. To address this issue, we introduce the safe exit approach.

4 Safe Exit Algorithm for Moving RkNN Queries and Moving Objects

In this section, we develop techniques to monitor the moving RkNN queries and moving objects in a road network. In Sect. 4.1 we present the algorithm for computing the safe region for moving q. Algorithm for computing safe exit points for UO is presented in Sect. 4.2.

4.1 Safe Region for Moving q

In this section, we present a new safe exit algorithm that addresses the issue for moving RkNN queries and moving objects in a road network. Algorithm 1 depicts the skeleton of our proposed safe exit algorithm for computing safe regions. It consists of three phases: (1) finding of useful objects that could contribute to the safe region, (2) computation of the influence region of the useful objects, (3) computation of safe exit points of query objects.

Algorithm 1: Computation of Safe Regions (skeleton)
Input: o: data objects, q: query object, k: number of requested RNNs
Output: SR: Safe Region
/*Phase 1: Retrieve useful objects*/
1:    object set $o' \leftarrow roadnetwork(o, q, k)$
2:    Object set $A \leftarrow \{o^+ \in o' | dist(o, q) < dist(o, o_{k+1})\}$
3:    Object set $NA \leftarrow \{o^- \in o' | dist(o, q) > dist(o, o_k)\}$
4:    **while** A or NA is non-empty **do**
5:      **Object** $o = pickobject(A, NA)$
6:      **If** $o \in A$
7:          $IR^+ \leftarrow computeIR(A, NA, k)$
8:          $SR = SR \cap IR^+$
9:      else
10:          $IR^- \leftarrow computeIR(A, NA, k)$
11:          $SR = SR - IR^-$
12:      **End while**
13:    **Return** SR

The above algorithm presents the skeleton of our proposed idea. Algorithm begins with finding the answer objects and non-answer objects. The detail methodology is explained in section Phase 1 section. Then in phase 2 it computes influence region of answer objects and non-answer objects. Next it computes the safe region by performing an intersection and set difference operations on road segments in Phase 3.

Phase 1: Finding Useful Objects

This phase aims at finding the potential objects from the network that can contribute to computation of safe regions. The goal is to retrieve the small set of data points in order to reduce the computation overhead. To achieve this we present pruning rules. Before we present pruning rules, we define the types of objects. The data objects are divided into two categories; namely, answer objects (denoted by O^+) and non-answer objects (denoted by O^-).

Definition 1 An object o is called an answer object if $sdist(o, q) < sdist(o, o')$ where o' is any other object in the road network.

Definition 2 An object o is called a non-answer object if $sdist(o, q) > sdist(o, o')$ where o' is any other object in the road network.

A simple method for retrieving R_k set is to traverse the network from q, and for each data object $o \in O$ encountered issue a nearest-neighbor query; if $q \in NN(o)$, which means q is the closest object to o. Consequently $o^+ \in R_k$. However, this approach needs to explore all the data objects since the size of R_k is not fixed and road network may contains points which are far from q. To avoid the unnecessary road network exploration we present the pruning Lemma. Before presenting Lemma, it is necessary to define closed nodes. A node n is called the closed node if there exists an object o such that $sdist(n, o) < sdist(n, q)$. The object o is called the blocking object because this is the object that makes node n a closed node. In Fig. 1, node n_2 is the closed node since $sdist(n_2, o_1) < sdist(n_2, q)$ which makes o_1 as the blocking object.

Lemma 1 *An object o cannot be the RNN of q if the shortest path between q and o contains a closed node with a blocking object o' where $o' \neq o$.*

Proof Let's assume that there exists a closed node n on the shortest path between o and q. The shortest distance between o and q is $sdist(o, q) = sdist(n, o) + sdist(n, q)$. Let o' be the blocking object and $sdist(o, o') = sdist(n, o) + sdist(n, o')$. As we know $sdist(n, o') < sdist(n, q)$, therefore, $sdist(o, o') < sdist(o, q)$. Hence, o cannot be the RNN of q.

In Fig. 1, the data object o_2 cannot be an RNN of q because the shortest distance between o_2 and q passes through n_2. Since $sdist(n_2, o_1) = 2$ and $sdist(n_2, q) = 3$ which makes data object o_1 more closer to o_2 than q.

Algorithm 2 illustrates the pseudo code for finding the answer objects. CORE traverses the network around q in a similar fashion to Dijkstra's algorithm and by using Lemma 1 it eliminates the nodes that may not lead to RNN. Algorithm begins by exploring the active edge where query object q is found. Each entry in the queue takes form $\langle p_{anchor}, edge \rangle$, where p_{ancor} indicates the anchor point in the edge. For active edge, q becomes the anchor point. Otherwise, either of the boundary node of the edge, i.e., n_s or n_e becomes the anchor point. If no desired number of answer objects found in an active edge, the edges adjacent to boundary nodes are en-queued. The traversal of edges is terminated when queue is exhausted. Line 4 initializes a *queue* by inserting the active edge. If edge contains a data object o, we need to verify if $o \in RNN(q)$. Thus algorithm, issues a *verify*(o, k, q) query (Line 10). The verification query checks weather q is among kNNs of data object o or not by applying a range-NN query around object o with range set to $sdist(o, q)$. If $q = kNN(o)$, which means o is the RkNN of q. Therefore o is added to the result set R_k (Line 12). If edge does not contain any data object o, the algorithm continues expansion and en-queues the adjacent edges of boundary node.

Algorithm 2: answer-object(q,k)
Input: q: query location, k: number of requested RNNs
Output: A_k: query result (answer objects)
```
1:        queue ← ∅       /* queue is a FIFO queue */
2:        A_k ← ∅         /* set of answer objects*/
3:        visited ← ∅     /*stores information of visited        edges */
4:        queue.push(q,edge_active) /* edge_active indicates active edge */
5:        while queue is not empty do
6:              ⟨p_anchor, edge⟩ ← queue.pop()
7:              if ⟨p_anchor, edge⟩ ∉ visited then
8:                    visited ←— visited ∪ {edge}
9:                          if edge contains a data object o
10:                             kNN(o): verify(o, k, q)
11:                          if q discovered by verification
12:                             R_k ← R_k ∪ o
13:                          Else
14:                             queue.push⟨β, edge⟩
15:              End while
16:              Return R_k
```

Next, we find the non-answer objects that can be useful to contribute the safe region. Useful non-answer objects $UO^- \in O^-$ are objects for which any of $o^+ = NN(o^-)$. In other words UO^- are RNNs of answer objects.

Pruning Rule 1: All answer objects are useful objects.

Explanation: We can generalize the above definition of answer objects to state that answer objects are RNNs. Therefore, all RNNs should be considered useful objects (UOs).

Pruning Rule 2: An object O cannot be a UO if its kNN does not contain any O^+ (answer object)

Explanation: From the definition of safe exit points (which we explain in detail in phase 3), the safe region includes the intersection of all answer objects and excludes the union of all non-answer objects.

In the given example, let's apply these pruning rules
2NN of $o_1 =(q, o_2) = (6, 7)$ | 2NN of $o_2 = (o_1, o_3) = (7, 8)$
2NN of $o_3 =(o_4, o_2) = (6, 8)$ | 2NN of $o_4 = (o_3, o_5) = (6, 10)$
2NN of $o_5 = (o_6, o_4) = (9, 10)$| 2NN of $o_6 = (q, o_5) = (8, 9)$

According to the pruning rules object o_1, o_2, o_5, and o_6 are useful objects (UO). Object o_3 and o_4 has been pruned.

Phase 2: Computing Influence Region for Useful Objects

After we retrieve the set of useful objects, the next step is to compute the influence region of answer and non-answer objects.

Influence Region of Answer Objects:

Influence region of answer objects is defined as:

$$I(o^+) = \{p | dist(o^+, p) \leq dist(o^+, o_{k+1})\}$$

Here o_{k+1} denotes the $(k+1)$th nearest neighbor of o. By definition the influence region of answer objects contains all the points for which $q = NN(o^+)$.

Influence region of answer objects can be computed by exploring the network around answer object in a similar manner as explained above in Sect. 4.1. The exploration terminates with the discovery of $(k + 1)$th nearest neighbor of answer object. The influence region will be marked by $range(o, d)$ where d is the $sdist(o, o_{k+1})$.

Figure 2 shows the influence region of object o_1 for example scenario discussed above. The expansion of road network starts from o_1 until it finds o_{k+1} which is 3NN in this example. Object o_4 is 3NN of o_1 and $sdist(o_1, o_4) = 8$. The algorithm will issue a $range(o_1, 8)$ query that marks all the points as influence region of o_1 as shows in Fig. 2. Similarly, influence region of o_4 can be computed.

Influence region of Non-answer Objects:

Influence region of non-answer objects is defined as:

$$I(o^-) = \{p | dist(p, q) < dist(p, o_k)\}$$

Influence region of non-answer objects can be computed from the distance of non-answer object to the kth object. Here o_k denotes the kth nearest neighbor of o. By definition the influence region of answer objects contains all the points for which $q \neq NN(o^-)$. In other words it contains all the points where object o remains o^-.

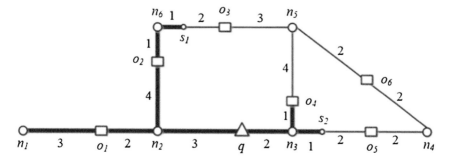

Fig. 2 Influence region of answer object o_1

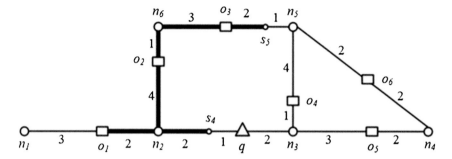

Fig. 3 Influence region of non-answer object o_2

Influence region of non-answer will be computed in same manner as Influence region of answer objects, the only difference is algorithm explores and computes distance to KNN object instead of $(k + 1)$NN.

Consider the object o_2 in Fig. 1:

2NNs of $o_2 = (o_1, o_3)$ with weights $(3, 6)$.

In this example, o_1 is the 2nd NN of o_2 and $sdist(o_2, o_1) = 6$. Therefore, Influence region of object o_2 is 6 units from o_2 to every connected direction. Similarly, if we consider o_5, its 2nd NN is o_4 and $sdist(o_2, o_4) = 4$. Thus, influence region of object o_5 is 4 units from o_5 to every connected direction. The bold lines in Fig. 3 shows the influence region of o_2. The NNs for answer objects changes when query object moves outside of influence region whereas in case of non-answer objects, the result changes when query object moves inside of influence region.

Phase 3: Computation of Safe Exit Points

The safe region $S(q, r)$ of a query "q" is defined as follows:

$$S(q, r) = \{\cap IR^+ - \cup IR^-\}$$

where IR^+ denotes the influence region of answer objects and IR^- denotes the influence region of non-answer objects. From definition we can see that any point that lies in the intersection of influence region of answer object O^+ regard as a safe region and any point p that lies in the influence region of non-answer object O^- should be excluded.

In a road network, the safe region is expressed by a set of segments. In Fig. 1, recall that o_1 and o_4 are answer objects and o_2 and o_5 are useful non-answer objects.

By applying above formula, the safe region can be expressed as

$$S(q, r) = \{IR(o_1) \cap IR(o_4) - IR(o_2) - IR(o_5)\}$$

Now let's compute the safe exit points for query object q in Fig. 1. In previous section, we already computed the influence region IR of all useful objects which described by set of segments. In order to obtain the safe region, we have to perform an intersection and set difference operations on road segments. We get segments $e(n_2, n_3)\}s\{(n_3, s_2), (n_3, o_4)\}$ by the intersection of IR(o_1) and IR(o_4). As o_2 and o_5 are non-answer objects, the safe region has to exclude the IR(o_2) and IR(o_5). Thus the safe region would be the segments (qs_4, qs_6). The set of boundary points s_4 and s_6 becomes the safe exit points.

4.2 Computation of Safe Exit Points for UO

In this section we compute the safe exit points for useful objects. As we are studying a special case where both data objects and query objects can move randomly in a road network, the RkNN for a query q can be changed by the motion of data objects. The safe region of query object q is constructed by UO objects. Therefore, instead of computing the safe exit points of all the objects, it is only necessary to compute the safe exit points of useful objects. This will significantly reduce the computation cost.

Recall that answer objects are those objects whose query object q is one of its kNN object. Which means answer object lies within the safe exit points until its kNN remains same. Similarly non-answer useful objects are those objects for which any of the answer object among its kNN. Which means all useful objects lies inside the safe exit points until their respective kNNs are same. Therefore, we need to monitor the nearest neighbor for moving objects in a road network. We are using the Cho et al. [4] Approach SEA that can efficiently computes the safe exit points of a moving nearest neighbor query on a road network.

First, we formally define a set of safe exit points for a moving NN query in the road network. Let p_{SE} be the set of safe exit points for a k-NN query point q and $O = \{o_1, o_2, \ldots, o_{|o|}\}$ be the set of objects of interest to q. Assume that the answer set (i.e., O^+) of q and its non-answer set (i.e., O^-) are $O^+ = \{o_1^+, o_2^+, \ldots, o_k^+\}$ and $O^- = \{o_{k+1}^-, o_{k+2}^-, \ldots, o_{|o|}^-\}$, respectively. Then, it holds that $d(q, o^+) \leq d(q, o^-)$ for an answer object $o^+ \in O^+$ and a non-answer object $o^- \in O^-$. Finally, p_{SE} is defined as follows:

$$P_{SE} = \Big\{ p_{se} \in G | MAX \left(d(p_{se}, o_1^+), d(p_{se}, o_2^+), \ldots, d(p_{se}, o_k^+)\right)$$
$$= MIN \left(d(p_{se}, o_{k+1}^-), d(p_{se}, o_{k+2}^-), \ldots d(p_{se}, o_{|0|}^-)\right) \Big\}$$

where MIN() and MAX() return the minimum and maximum values of the input array, respectively. In other words, a safe exit point pse is the midpoint (i.e., $MAX(d(p_{se}, o_1^+), \ldots, d(p_{se}, o_k^+)) = MIN\left(d(p_{se}, o_{k+1}^-), \ldots, d(p_{se}, o_{|0|}^-)\right)$) between the farthest answer object and the nearest non-answer object.

The following two main lemmas have been presented in paper to decide whether safe exit point exists in the segment or not.

Lemma 2 *If $A_\beta \cup O_{(n_\beta, p_{anchor})} \neq A_{p_{anchor}}$, there is a safe exit point p_{se} in the segment.*

Proof Please refer to [4]. This lemma means the safe exit point in a segment exists if set of answer objects at n_b is not equal set of answer objects at p_{anchor}.

Lemma 3 *If $A_\beta \cup O_{(n_\beta, p_{anchor})} = A_{p_{anchor}}$, there is no safe exit point in the segment.*

Proof Please refer to [4]. This lemma means the safe exit point in a segment does not exists if set of answer objects at n_b is equal set of answer objects at p_{anchor}.

We now discuss the computation of the safe exit points for the answer object o_1 in the example road network shown in Fig. 1. Recall that we are considering $k = 2$. The answer objects of data object o_1, $A_{o_1} = \{q, o_2\}$. SEA starts exploration from the active edge where data object o_1 lies. Since $e(n_1, n_2)$ is the active sequence, the location of o_1 is the anchor point. Each of the two segments $s(n_1, o_1)$ and $s(o_1, n_2)$ within $e(n_1, n_2)$ is explored individually. For $s(n_1, o_1)$, $p_{anchor} = o_1$, $A_{o_1} = \{q, o_2\}$, $A_{n_1} = \{q, o_2\}$ and $O_{(n_1, o_1)} = \{\emptyset\}$. By Lemma 3 $A_{n_1} \cup O_{(n_1, o_1)}$ there is no safe exit point within $s(n_1, o_1)$. Similarly, for segment $s(o_1, n_2)$ there is no safe exit point by Lemma 4.

Therefore, edges adjacent to n_2 are explored with $n_2 = p_{anchor}$. The edge $e(n_2, n_3)$ will be explored next. For $e(n_1, n_3)$, $p_{anchor} = n_2$, $A_{n_2} = \{q, o_2\}$, $A_{n_3} = \{q, o_4\}$ and $O_{(n_2, n_3)} = \{q\}$. By Lemma 2 i.e., $A_{n_3} \cup O_{(n_2, n_3)} \neq A_{n_2}$ a safe exit point exists in the edge. For each point $p \in e(n_2, n_3)$, $o^+_{farthest}$ will be selected from the answer objects in $A_{n_2} = \{q, o_2\}$ while $o^-_{nearest}$ will be selected from the non-answer objects in $A_{n_3} \cup O_{(n_2, n_3)} - A_{n_2} = \{o_4, o_5\}$. As shown in Fig. 4, $o^+_{farthest} = o_2$ because for every point $p \in e(n_2, n_3)$, $dist(p, o_2) > dist(p, q)$ while $o^-_{nearest} = o_4$ because for every point $p \in e(n_2, n_3)$, $dist(p, o_4) < dist(p, o_5)$. The safe exit point p_{se1} is the midpoint between o_2 and o_4. That is $dist(p_{se1}, o_2) = dist(p_{se1}, o_4)$ where $dist(p_{se1}, o_2) = x + 4$ and $dist(p_{se1}, o_4) = -x + 6$ for $0 < x < 5$. Consequently, $x = 1$. This means that the distance from n_2 to p_{se1} is 1.

Similarly, safe exit point p_{se2} in the edge $e(n_2, n_6)$ can be determined.

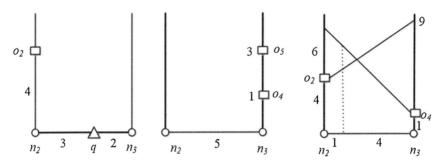

Fig. 4 Determination of safe exit point p_{se1}

5 Performance Evaluation

In this section, we describe the performance evaluation of proposed algorithm CORE by means of simulation experiments. Since traditional naïve algorithm recomputes the result at every time stamp, therefore, instead of comparing CORE with it, we made it more competitive by applying our approach with zero safe regions and called it Naive$_u$. Now this Naive$_u$ recomputes the query result only when the query object or data object changes its location instead of every timestamp. Consequently, it improves the performance of naïve algorithm.

5.1 Experimental Settings

All of our experiments are performed using a real road network [9] for San Joaquin County, California, USA, which contains 18,263 nodes and 23,874 edges. We simulate moving objects (query as well as data objects) using the network based moving objects generator [2]. Table 2 lists the default parameters used in our experiments.

5.2 Experimental Results

Firstly, we study the effect of number of objects in the performance of Naive$_u$ and CORE algorithm. Figure 5 shows the performance of Naive$_u$ and CORE w.r.t number of objects. Both of the algorithms are sensitive to the increase in number of objects because the algorithm has to handle the updates of more objects. However, CORE clearly outperforms Naive$_u$.

In Fig. 6, we study the effect of number of objects on communication cost. It shows that the messages sent by both algorithms to server tended to increase as the number of objects increased. However, CORE shows better performance because of the fact that when the query and data objects remains within the safe exit points, recomputation of query results is not required, which ultimately reduces the number of messages sent between query and server.

Parameter	Range
Number of objects	1, 5, **50**, 70, 100 ($\times 1000$)
Number of queries (N_{qry})	1, 3, **5**, 7, 10 ($\times 1000$)
Number of requested RNNs (k)	8, 16, **32**, 64, 128

Table 2 Experimental parameter settings

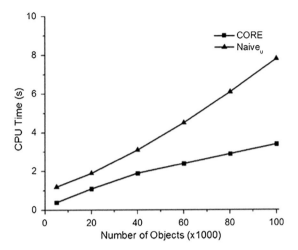

Fig. 5 Effect of number of objects on computation cost

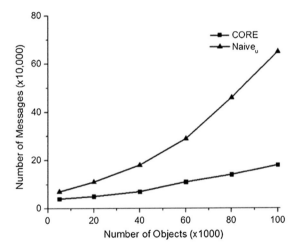

Fig. 6 Effect of number of objects on communication cost

Figure 7, shows the performance trends of CORE and the Naive$_u$ algorithm w.r.t. the number of queries. Experimental results revealed that the computation time of both algorithms increased as the number of queries increased. The computation time of CORE increases mainly because with the increase of queries more useful objects needed to be found. Consequently, algorithm required to compute the p_{se} of more objects which degrade the performance.

Fig. 7 Effect of number of queries on computation cost

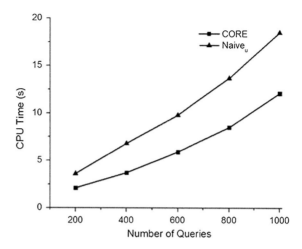

6 Conclusion

In this paper, we studied the processing of continuous RkNN queries in road networks where both query and data objects are moving and introduced a new algorithm called CORE. Our approach is based on the safe exit points which can significantly improve not only the computation cost but also communication cost between server and query object. The results of experiments conducted using real datasets indicate that our algorithm drastically reduces computation cost and communication cost compared to Naive$_u$ algorithm. This study can be further extended to directed road networks or privacy aware systems where the location of query is hidden from the server.

Acknowledgments The research was supported by Basic Science Research Program through the National Research Foundation of Korea (NRF) funded by the Ministry of Education (2013R1A1A2A 10012956, 2013R1A1A2061390).

References

1. Benetis, R., Jensen, C., Karciauskas, G., Saltenis, S.: Nearest neighbor and reverse nearest neighbor queries for moving objects. In: IDEAS, pp. 44–53 (2002)
2. Brinkhoff, T.: A framework for generating network-based moving objects. Geo Inform. **6**(2), 153–180 (2002)
3. Cheema, M., Lin, X., Zhang, Y., Zhang, W., Li, X.: Continuous reverse k nearest neighbors queries in Euclidean space and in spatial networks. VLDB J. **21**, 69–95 (2012)
4. Cho, H., Kwon, S., Chung, T.: A safe exit algorithm for continuous nearest neighbor monitoring in road networks. Mob. Inf. Syst. **9**(1), 37–53 (2013)
5. Cho, H., Ryu, K., Chung, T.: An efficient algorithm for computing safe exit points of moving range queries in directed road networks. Inf. Syst. **41**, 1–19 (2014)
6. Kang, J., Mokbel, M., Shekhar, S., Xia, T., Zhang, D.: Continuous evaluation of monochromatic and bichromatic reverse nearest neighbors. In: ICDE (2007)

7. Korn, F., Muthukrishnan, S.: Influence sets based on reverse nearest neighbor queries. In: SIGMOD (2000)
8. Li, G., Li, Y., Li, J., Shu, L., Yang, F.: Continuous reverse k nearest neighbor monitoring on moving objects in road networks. In: Information Systems (2010)
9. Real datasets for spatial databases, http://www.cs.fsu.edu/~lifeifei/SpatialDataset.htm
10. Safar, M., Ebrahimi, D., Taniar, D.: Voronoi-based reverse nearest neighbor query processing on spatial networks. In: Multimedia Systems (2009)
11. Sun, H., Jiang, C., Liu, J., Sun, L.: Continuous reverse nearest neighbor queries on moving objects in road networks. In: WAIM, pp. 238–245 (2008)
12. Wang, H., Zimmermann, R.: Processing of continuous location-based range queries on moving objects in road networks. IEEE Trans. Knowl. Data Eng. **23**(7), 1065–1078 (2011)
13. Yiu, M., Mamoulis, N., Papadias, D., Tao, Y.: Reverse nearest neighbor in large graphs. In: ICDE, pp. 186–187 (2005)

A Voice Dialog Editor Based on Finite State Transducer Using Composite State for Tablet Devices

Keitaro Wakabayashi, Daisuke Yamamoto and Naohisa Takahashi

Abstract In recent times, mobile spoken-dialog systems have become increasingly widespread. We have developed MMDAgent, a toolkit for building voice interaction systems. Users can customize MMDAgent to their liking by writing Finite State Transducer (FST) scripts. However, we discovered that beginners find it difficult to edit FST scripts manually because the number of states become larger to describe natural dialog. To resolve these problems, in this paper, we propose a method of editing voice interaction contents using composite state. The results of experiments conducted indicate that this objective was achieved.

1 Introduction

Mobile spoken-dialog systems (exemplified by Siri, from Apple Inc. [1].) have become increasingly widespread in recent years. We studied mobile spoken-dialog systems and subsequently developed MMDAgent [2, 3], a toolkit for building voice interaction system. MMDAgent is able to run on the Android platform [4]. MMDAgent is a spoken-dialog system that enables users to converse with a 3D character. It is currently utilized in an interactive spoken guidance digital signage system [5] at the main entrance to Nagoya Institute of Technology.

MMDAgent enables user to edit voice interaction scenarios by editing a file called a Finite State Transducer (FST) script, which it loads and executes on startup. User need to edit the FST script in the proper format in order to customize the voice

K. Wakabayashi (✉) · D. Yamamoto (✉) · N. Takahashi (✉)
Nagoya Institute of Technology, Gokiso-cho, Showa-ku,
Nagoya-shi, Aichi 466-8555, Japan
e-mail: wakabayashi@moss.elcom.nitech.ac.jp

D. Yamamoto
e-mail: yamamoto.daisuke@nitech.ac.jp

N. Takahashi
e-mail: naohisa@nitech.ac.jp

© Springer International Publishing Switzerland 2016
R. Lee (ed.), *Computer and Information Science 2015*,
Studies in Computational Intelligence 614, DOI 10.1007/978-3-319-23467-0_9

interaction scenario. When a system can manage various situations, end-users feel it is user-friendly. However, to achieve this, users need to add many states to handle various situations.

The advantage offered by an FST script is that it enables user to describe complex dialogs containing voice, pictures, motions, and expressions in accordance with their desires. However, the script has the drawback that many complex scripts are required in order to describe complex dialogs. User find it time-consuming and difficult to read such state transition diagrams.

In general, in order to describe rich voice interaction, a combination comprising not only words, but also tone of voice, motion, expression, and pictures is desired. However, describing such a combination in the format used by an FST script is complicated and difficult for the average user. Consequently, we propose a method for making templates comprising frequently used elements of complex dialogs that users can utilize to generates elements of rich dialogs more easily.

Further, in order to treat these elements as inner transitions of composite states, we propose a method of generating large dialogs by combining elements of such dialogs. In particular, as a feature of voice interaction, composite states are connected to each other by transitions, whose transition conditions are external events, because dialogs branch in response to external events, for example voice inputs form users.

This composite state is different from composite state of UML state diagrams. The composite state of UML state diagrams shows internal state. In contrast, this composite state does not show internal state because the purpose of this feature is encapsulation.

The proposed method has the following features:

1. Users can edit state diagrams using a GUI.
2. Composite states are used in order to reduce the number of superficial states.
3. Templates of transducer patterns used frequently are made in advance.

This paper discusses this proposed system, which makes is easy to edit voice interaction scenarios for MMDAgent for Android using tablets, regardless of the number of states comprising each scenario.

2 MMDAgent

2.1 Overview of MMDAgent

MMDAgent incorporates low-latency fast speech recognition, speech synthesis, embodied 3-D agent rendering with simulated physics, and dialog management. It enables user to converse with 3-D character on screen such as that depicted in Fig. 1. The speech recognition engine utilized is Julius, the speech synthesis engine is OpenJTalk, and the 3-D model and motion file format is MikuMikuDance format.

Fig. 1 MMDAgent for Android

A major feature of MMDAgent is that it allows users to customize voice interaction scenario in accordance with their desires. The voice interaction scenarios format utilized is FST—a kind of Finite State Automaton. As stated above, voice interaction scenarios are written in an "FST script," and customized by editing it.

2.2 FST Script

Figure 2 shows an example of an FST script.

As Fig. 2, an FST script consists of rows expressing transitions. A row consists of source state number, destination state number, transition condition (event), and command. The top half of Fig. 2 describes a scenario for "Hello." It states that when the system hears/recognizes the word "Hello" in the user's voice, then Mci (3-D character) should say "Hello" and also bow. The state with the number 0 is the start state in the FST script.

2.3 FST Script Problems

Editing FST script is difficult for beginners because the number of states quite large in cases where the user wants to handle many situations in the description of natural dialog. An FST script is non-intuitive because it describes state diagram in text format. Consequently, understanding how to transit by reading FST scripts is difficult. Additionally, it is possible to write transitions that have the same state but at distant locations in FST script. User cannot avoid such a description in cases where the

```
# Sample pattern 1: Hello
  0   10   RECOG_EVENT_STOP|hello                  SYNTH_START|mei|normal|Hello.
 10   11   <eps>                                   MOTION_ADD|mei|action|Motion\greeting.vmd|PART|ONCE
 11    0   SYNTH_EVENT_STOP|mei                    <eps>

# Sample patte┌ Destination state number ┐          ┌ Command ┐
  0   20   RECOG EVENT STOP|self,introduction  SNTH START|mei|normal|My name is Mei.
  0   20   RECOG EVENT STOP|who you             SYNTH START|mei|normal|My name is Mei.
 20   └ Souce state number ┘        └ Transition condition ┘ADD|mei|action|Motion\self_introduction.vmd|PART|ONCE
 21   22   SYNTH_EVENT_STOP|mei                 SYNTH_START|mei|normal|Nice to meet you.
```

Fig. 2 Example of an FST script

Fig. 3 State diagram
describing Sample Pattern 1
in Fig. 2

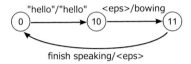

transducer is complex. In cases where a transition is far from other transitions that
have same state, understanding how to transit is difficult.

Moreover, conflicting state number is a common bug that occurs while editing
FST script. When a user adds a state while editing an FST script, the user has to
check that the number intended for use as the new state is not already in use. When
the FST script has many state, users often miss conflicting state numbers.

A translation of the dialog from Sample Pattern 1 in Fig. 2 is shown as a state
diagram in Fig. 3. The state diagram makes it easier to understand the state transducer
intuitively in this visual form rather than as text.

The greater the number of states is, the greater the number of elements. In such a
case, the screen becomes complex and it is difficult to operate even when the state
transducer is described visually. Therefore, a method that is easy to operate, even
when the number of state is large, is necessary.

3 The Proposed Method

3.1 Overview of the Proposal

In this paper, we propose a method of editing voice interaction contents using composite state. Specially, we propose a system that uses MMDAgent for Android and
runs on the Android OS.

The merit of editing voice interaction scenarios visually is easy to understand state
transducer intuitively. Moreover, using composite states and integrating states into
meaningful groups, facilitates the editing of state transducers from macro viewpoint.
Consequently, operability becomes better because the number of states decreases at
the same time. Features of the proposed method are as follows:

1. User can edit state diagram using a GUI.
2. Composite states are used in order to reduce the number of superficial states.
3. Template of transducer patterns used frequently is made in advance.

3.2 Model of State Diagram Using Composite State

FST scripts use Mealy Machines as transducer model. Using a Mealy Machine model, it is difficult to see the entire diagram as the number of states explode. Therefore, we propose a transducer model that integrates composite state to Mealy Machine model.

Integrating composite state enables the formation of meaningful groups. Consequently, the number of states decreases and the diagram become editable from a macro viewpoint.

Figure 4 shows the state transducer model proposed. The model is composed of atomic states, composite state, and transitions as the figure shows. A composite state is a state comprised of internal states, a state that is not composite state is an atomic state, and a transition is a directed edge that connects two states.

Figure 5 gives an example of the proposed state transducer model. As shown in the figure, the state diagram has a tree structure. Rectangles represent composite states, and circles represent atomic states. An arrow shows that a destination state is the inner state of source state.

The state root, S, and M is managed by the system; users can edit only states under state M.

Introducing composite state, it is enable to manage states as meaningful group.

Fig. 4 Example of the state transducer model

Fig. 5 Example data structure of the state diagram

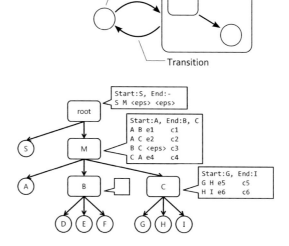

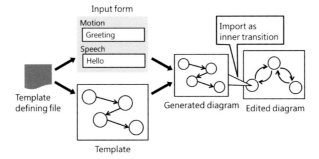

Fig. 6 Example illustrating how an FST template is utilized

3.3 FST Template

The proposal system has a function that enables editing of state diagrams using fewer operation than the conventional system. This function utilizes transducer pattern templates. Using FST templates makes it easy to edit state diagrams that have complex transitions or are used frequently.

An FST template and its input form is prepared in advance. They defined in an external file. Text box and pull-down list are available for use in input form.

Figure 6 shows an example of the generating state diagrams from FST template. As shown int the figure, the content of the state transition is obtained from the FST template and input contents. The generated state transition is then added to state diagram as a composite state.

3.4 Translating from Conventional Model to the Proposal Model

In order to load interaction scenarios into MMDAgent, an FST script must be generated. However, a conventional FST script cannot contain composite states. Therefore, a translating state diagram is necessary. The translation process utilizes the following steps:

1. Set a number to the atomic states.
2. Expand composite states.
3. Export as FST script.

From this point onwards, we explain the translation process using the data shown in Fig. 5 as an example.

Fig. 7 Diagram condition
after setting number to the
states in Fig. 5

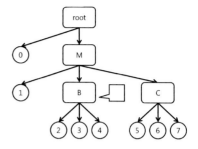

3.4.1 Setting a Number to the Atomic State

FST scripts identify states by numbers; therefore, the system set a number to an
atomic state that has no duplicate. The process of setting the number is as follows:

1. Set zero to state S.
2. Search nodes recursively from state M.
3. Set the minimum number in the set of numbers that has not been set to a state,
   if the node is atomic state.

State S must be set the number 0 because the state having the number 0 is always
the initial state in an FST script. State S is set to zero because it is initial state in the
interaction scenario edited using proposal system. Figure 7 shows the condition of
the diagram after setting number to states in Fig. 5.

3.4.2 Spreading Composite States

An FST script's state transition model dose not have composite states. Therefore,
composite states must be expanded during the conversion to FST script. The expansion process is in essence the same as state machine diagram in UML. For example,
when a state diagram such as that depicted in Fig. 8a is expanded, transitions going
out to composite states are reconnected to all inner states, as depicted in Fig. 8b.

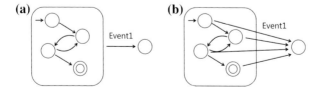

Fig. 8 Expanding a composite state. **a** Before. **b** After

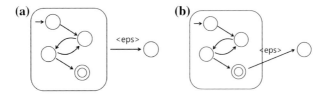

Fig. 9 Expanding a composite state (with $< eps >$ transition). **a** Before. **b** After

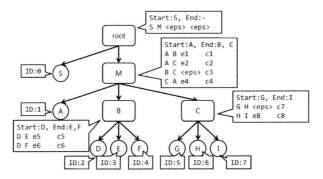

Fig. 10 State diagram data before exporting

However, if a state diagram such as that shown in Fig. 9a is expanded, the transition with the $< eps >$ condition is reconnected to the final states of the composite state, as illustrated in Fig. 9b.

In FST script, if a state is set a transition with the $< eps >$ condition, the state immediately change to next state. Thus, the transitions are reconnected to all the inner states during the expansion and the composite state is changed to the next state of the composite state. Therefore, the meaning of the $< eps >$ condition in the proposal system is "transit if it is in final state," and $< eps >$ transition is reconnected to final states of the composite state.

3.4.3 Exporting as FST Script

Finally, the interaction scenario is exported as FST script to be loaded into MMDA-gent. The state diagram with numbers set, as shown Fig. 10, is then exported and converted to the FST script shown in Fig. 11.

Fig. 11 FST script
generated from the state
diagram in Fig. 10

```
0   1   <eps>   <eps>

1   2   e1      c1
1   5   e2      c2
3   5   <eps>   c3
4   5   <eps>   c3
5   1   e4      c4
6   1   e4      c4
7   1   e4      c4

2   3   e5      c5
2   4   e6      c6

5   6   <eps>   c7
6   7   e8      c8
```

4 Prototype

We implemented a prototype system that enabled us to edit voice interaction scenarios using the proposal method. The prototype system was implemented using the Android SDK and executed on an Android OS device.

4.1 Flow of Editing

The procedural flow of editing of the FST script using this prototype system was as follows:

1. Run prototype system.
2. Add state or transition to diagram.
3. Input substitute to state or transition.
4. Repeat steps 2 and 3.
5. Tap run button and export FST script.

4.2 Interface

4.2.1 Main Screen

Figure 12 shows the main screen of the system. A canvas is on the left of the screen and the MMDAgent view on the right. The MMDAgent view is used to try edited scenarios and operate the system using voice.

Fig. 12 Main screen

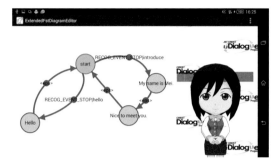

Fig. 13 State setting dialog

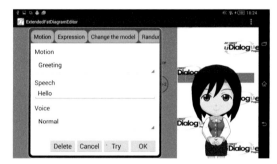

4.2.2 State Setting Dialog

When state is tapped before the "setting" menu is selected, the state setting dialog appears as Fig. 13. Selecting the template using the button at the top of dialog results in the input form at the bottom of dialog changing. The actions in the setting can then be tested by tapping the "Try" button.

4.2.3 Transition Setting Dialog

When transition is tapped before the "setting" menu is selected, the transition setting dialog in Fig. 14 appears. The keyword that generates the transition can be entered either by keyboard or voice.

4.3 Functions of the Prototype

The prototype system has the following functions:

1. FST template.
2. Keyword entry via voice.
3. Adjustment of the position of the arrows.

Fig. 14 Transition setting dialog

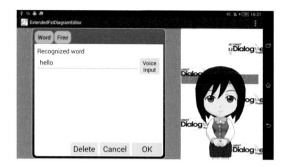

4.3.1 FST Template

The FST template is a stored procedure that is used frequently, it is prepared in advance. The system generates transitions after the user selects a template similar to the desired procedure and entering the relevant data.

4.3.2 Keyword Entry via Voice

Voice input reduces the need for keyboard input. The voice recognition engine used is the same engine used by MMDAgent; thus, recognition error is minimized.

4.3.3 Adjustment of the Position of the Arrows

When arrows are being added, the system adjusts the position of arrows to ensure that they do not overlap. Thus, it is able to create and edit a readable diagram.

5 Experiments and Considerations

5.1 Usability Experiment

The purpose of this experiment was verification that the system resolves the problems encountered using FST scripts in the conventional system and makes it easy to edit scenarios.

The experiment was conducted as follows. Eight students from the university participated as subjects. The subjects edited scenarios as requested using both the conventional and the proposed methods, and the time taken to complete each measured. They then completed questionnaires after the experiment that allowed us to compare the two methods. The questionnaires were five-rated and included space for comments.

The two methods listed below were utilized for the editing process. One-half of the subjects edited using method 1 before method 2, and the other edited in the reverse order:

1. Edit FST script using FstFileEditor [6].
2. Edit FST script via diagrams by using the prototype.

We decided to utilize keywords and reaction (speech and motion) in the editing scenarios.

Table 1 lists the items used in the questionnaire. Both methods were evaluated on each of the items. The evaluation was five-point rated. as follows: 1: yes, 2: maybe yes, 3: not sure, 4: maybe no, and 5: no.

Figure 15 shows the average value obtained from evaluation of the questionnaires. Larger values indicate better performance in items 1 through 12, whereas the reverse holds true for items 13 to 15.

In method 1, almost all items have better values than those in method 2. The values for items 1 through 6 are high (almost 4.5 points). However, item 3 ("For beginner") in method 1 has an approximate values of 1.5 point, thus, method 1 is not for beginners. Conversely, this result indicate that the proposed system is effective for beginners.

Table 1 Items in questionnaires

Number	Question
1	Fun to use
2	Easy to edit
3	For beginner
4	Want to use again
5	Able to edit intuitively
6	Easy to understand how you connect transition
7	Easy to understand transition
8	Easy to remember how to edit
9	Able to edit as desired
10	Able to edit high-quality scenarios
11	Able to edit complex scenarios
12	Bug cannot occur easily
13	Easy to miss
14	Tiring to operate
15	Frustrating to operate

Fig. 15 Results: Lager values are better for items 1 through 12; smaller values are better for items 13 through 15

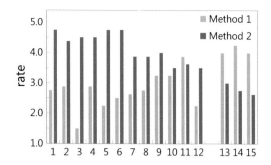

Table 2 Average time to edit (s)

	Method 1	Method 2
Group A	1313	707
Group B	1263	770
Both group	1288	739

The value for item 12 in method 2 is higher (3.5 points) than the corresponding value in method 1 (2.3 points). Thus, the previous problem of bugs occurring easily has been resolved. Additionally, Table 2 shows the average time taken to edit the scripts. According to Table 2, method 2 becomes 449 s faster than method 1. (The time taken to edit using method 1 is 1288 s, compared to 739 s for method 2.)

5.2 Comparison of Sum of State of Entrance Signage

Table 3 compares the number of states used to represent the entrance signage in methods 1 and 2. The number of states used in the template to the number of states in methods 1, because each composite state has two extra states (initial state and final state). The number of states used in the template was 200.

As shown Table 3, the superficial number of states was one-tenth that of method 1. However, The number of states in method 2 changed with prepared templates.

Table 3 Comparison of sum of state of entrance signage

	Method 1	Method 2	
		Superficial	Substantial
Sum of states	2729	257	3200

6 Related Systems

A few systems, such as Islay [7] and MMDAE [8], are related to this system. Islay is a tool for editing interactive animation using state diagrams. Event-driven animation can be created to edit state diagram constructed using Moor machines. It is related to our system in that it also uses visual programming for state diagram. However, our system is targeted at creating voice interaction contents, whereas Islay's target is the creation of interactive animation.

MMDAE is an FST script editor that runs on web browser. It has events, commands, and arguments complement functions. It has a feature that changes the interface to adjust to the user's skill level. It is related to our system in that the editor utilizes voice interaction for contents, as does MMDAgent. However, our system uses GUI-based state diagram, whereas MMDAE is text-based.

7 Conclusion

In this paper, we proposed a voice dialog editor using composite state for tablet devices. We also implemented a prototype system and conducted experiments.

The proposal method has the following features: (1) User can edit state diagram using a GUI. (2) Composite states are used in order to reduce the number of superficial states. (3) Template of transducer patterns used frequently is made in advance.

The results of comparative experiments conducted using the conventional and proposed methods indicate the following. The proposed method (score: 4.8 points) is more fun to use than the conventional method (score: 2.8 points), where a higher score is better. The proposed method (score: 4.4 points) is easier to edit than the conventional method (score: 2.9 points). The proposed method is also intuitively easier to edit (score: 4.8 points) than the conventional method (score: 2.3 points). Further, the proposed method is more effective for beginners.

In future work, we plan to provide a scenario edited by this system. In addition, we plan to provide a library and generalize the functions of the FST template because combining template and data to generate substantive transitions is effective in other systems for generating FST scripts.

Acknowledgments This research was supported by CREST, JST.

References

1. Apple Inc., Siri, www.apple.com/ios/siri/. Accessed 10 May 2015
2. Lee, A., Oura, K., Tokuda, K.: In: 2013 IEEE Int'l Conference on Acoustics, Speech and Signal Processing (ICASSP) (2013), pp. 8382–8385. doi:10.1109/ICASSP.2013.6639300
3. MMDAgent, www.mmdagent.jp. Accessed 10 May 2015

4. Yamamoto, D., Oura, K., Nishimura, R., Uchiya, T., Lee, A., Takumi, I., Tokuda, K.: In: Proceedings of the 2nd International Conference on Human-agent Interaction (HAI'14, ACM, New York, 2014), pp. 323–330. doi:10.1145/2658861.2658874
5. Oura, K., Yamamoto, D., Takumi, I., Lee, A., Tokuda, K.: J. Jpn. Soc. Artif. Intell. **28**(1), 60 (2013)
6. negi. Fstfileeditorver2.01. Negi, FstFileEditorVer2.01, http://d.hatena.ne.jp/CST_negi/20121 215/1355564598. Accessed 10 May 2015
7. Okamoto, S., Kamada, M., Nakao, T.: IPSJ J. Program. **46**(1), 19 (2005)
8. Nishimura, R., Yamamoto, D., Uchiya, T., Takumi, I.: In: Proceedings of the 2nd International Conference on Human-agent Interaction, HAI'14 (2014), pp. 129–132. doi:10.1145/2658861. 2658904

Analysis of Driving Behaviors Based on GMM by Using Driving Simulator with Navigation Plugin

Naoto Mukai

Abstract Generally, novice drivers often feel uneasy about driving in unfamiliar traffic environment. In Japan, the road traffic act was revised in November, 2011. This revision authorized the operation of the roundabout intersection which is a traffic-lights-free intersection. The roundabout intersection have been widely spread in Europe, on thes other hand, Japanese people are unfamiliar with the roundabout intersection, and especially novice drivers may panic easily. Thus, this paper aims for modeling driving behaviors to support operation of novice drivers. Moreover, we examine the effects of navigation at the roundabout intersection for the novice drivers. First, we built a driving simulation environment for roundabout intersection by using UC-win/Road, and we collect driving logs of a novice driver who drives with navigation or without navigation. Then, we classify the driving log into some discrete driving states (i.e., go straights) on the basis of Gaussian Mixture Model (GMM) and plot state transition graphs of the discrete driving states to model driver's behavior. The driver model obtained by above method showed the characteristic driving behaviors of novice drivers in the simulation environment.

1 Introduction

In these years, many kinds of sensor devices for vehicles are in widespread use. Such sensor devices provide various driving information (e.g., GPS, speed, and acceleration) easily. Moreover, we can obtain further information related to driving behaviors (e.g., steering anglc, pedal pressure, and so on) from analysis of ECU (Electronic Control Unit) and in-vehicle network called CAN (Controller Area Network). Hiroki et al. retrieved driving information from ECU via ELM327 which is one of the on-board diagnostics scan tools, and suggested three evaluation standards related to safety and so on [11]. Ishikawa et al. proposed a method of identifying individuals

N. Mukai (✉)
Department of Culture-Information Studies, School of Culture-Information Studies,
Sugiyama Jogakuen University, 17-3, Hoshigaoka-motomachi, Chikusa-ku, Nagoya,
Aichi 464-8662, Japan
e-mail: nmukai@sugiyama-u.ac.jp

© Springer International Publishing Switzerland 2016
R. Lee (ed.), *Computer and Information Science 2015*,
Studies in Computational Intelligence 614, DOI 10.1007/978-3-319-23467-0_10

based on the driving information from CAN [2, 7]. They also reported that the identification rate by using the pedal pressure is very high. As above, it is obvious that the sensor data of vehicles is important factor to improve our driving environment.

In Japan, roundabout intersections have received a lot of attention. The roundabout is a traffic-lights-free intersection which have been frequently used in the Europe. According to the revision of Japanese traffic act in November, 2011, the roundabout intersection can be introduced into actual road traffic in a formally way. However, most of Japanese people are unfamiliar with the roundabout intersection, especially novice drivers may panic easily at the place. Previously, some technical researches about the roundabout intersection were reported. Hasegawa et al. evaluated the comparison between roundabout intersections with signalized intersection [10], they indicated that the roundabout intersection is effective for off-peak roads. The consideration of these paper is very useful for city planning, but we think that the support for novice or older drivers [6, 8] is also important to produce the best performance of roundabout intersections.

Thus, we first designed a virtual driving course by using UC-win/Road which is a driving simulator software developed by Forum8 Co., Ltd. (http://www.forum8. com). The virtual driving way is based on an actual roundabout intersection located at Ichinomiya, Aichi in Japan. In order to evaluate the effect of navigation at the roundabout intersection for novice drivers, we developed a navigation plugin which displays texts and images on the screen. Then, we analyze driving logs of a novice driver by using Gaussian Mixture Model (GMM). The GMM is one of the popular methods for modeling drivers. For example, Mima et al. proposed a GMM based method to estimate the driving state of drivers by using time-series data of brake pressure [9]. Their results showed that the method can discriminate between the brake for waiting at the traffic light and the brake for turning right or left. In this paper, we pick up accelerator and brake pressures as time-series data and classify the driving states into some categories. Moreover, we compare state transition graphs based on the some categories to identity key difference between novice drivers with navigation and without navigation.

The construction of this paper is as follows. Section 2 explains the virtual driving course we designed on the basis an actual roundabout intersection. Section 3 shows the original driving log which is obtained by a driving simulator software UC-win/Road. We describe the clustering process based on the GMM in Sect. 4, and compare transition graphs which imply the characteristics of drivers in Sect. 5. Finally, we conclude this paper in Sect. 6.

2 Driving Simulation

The MLIT (Ministry of Land, Infrastructure, Transport and Tourism) reported that there are about 140 roundabout intersections in Japan (http://www.mlit.go.jp/road/ir/ ir-council/roundabout/). The introduction of roundabout intersections will become popular with the enforcement of the revised traffic act. In order to support novice drivers at the roundabout intersection, we developed a virtual driving course by using UC-win/Road which is one of the famous driving simulator software. The software has been widely used for various research fields [1, 5, 12]. The detail of the software is opened on the official website (http://www.forum8.com).

Here, we pick up a roundabout intersection which is located in Ichinomiya, Aichi in Japan. Figure 1 is a picture of the roundabout intersection at Ichinomiya. This intersection is classified into one-lane roundabout, and the perimeter is about 40 m. In roundabout intersections, vehicles drive round the circle in a clockwise direction, and vehicles in the circler road have a priority to other incoming vehicles. The virtual driving course based on the roundabout intersection at Ichinomiya is shown in Fig. 2. There are two kinds of intersections along the main street in the course. A test driver encounters a normal signalized intersection at first, and then the test driver encounters a one-lane roundabout intersection. After that, the test driver follows the instructions and returns to the original location. Moreover, the test driver receives the predetermined instructions at the eight spots in the course. For example, a text

Fig. 1 Roundabout intersection at Ichinomiya

Fig. 2 Virtual driving course and instruction spots

Table 1 Instruction for driving

Spot number	Instruction text
1	Go straight
2	Stop here
3	Take the third exit
4	Exit here
5	Turn left
6	Turn left
7	Keep straight
8	Turn left

"Go Straight" and relevant image are displayed on the screen at the spot no.1. Other instructions are summarized in Table 1. This function is realized by a navigation plugin we developed.

3 Original Driving Log

We adopted some students in our university as test drivers for the virtual driving course. Most of them are short on experience in driving and feel like bad drivers themselves although they have driver's license. The experimental results of test driving showed that they have similar characteristics unique to novice drivers. In addition, the feedback after the driving indicated that they felt that the roundabout intersection is fear and difficult. Here, we pick up one test driver from them to discuss the effect of the navigation plugin in greater depth. The driving logs we deal with are two categories: accelerator and brake pedals, and each test driver drives the course twice: without and with navigation plugin. Consequently, we compare four patterns of driving log summarized in Table 2. The four patterns are denoted PT1, PT2, PT3, and PT4 for simplicity in this paper. The original driving logs of the four patterns are graphed in Figs. 3, 4, 5, and 6. These logs can be recorded by original log function of UC-win/Road. The horizontal axis of the graphs shows elapsed time (second) from the start. The driving time was reduced from 4139 to 2555 s by using the navigation plugin. The vertical axis of the graphs shows the pressure rate of accelerator or brake pedals. The value range of the pressure rate is from zero to one: zero means that the pedal is completely released, and one means that the pedal is pressed down strongly. It seems that the graphs indicate that the navigation plugin can support to smooth driving because the amplitude of the signal is relatively small and little. We note the fact that the experience of first driving may affect the driving log of second driving. However, the feedback after driving also indicated that they felt that the driving with

Table 2 Comparison pattern of driving log

Pattern No.	Log	Navigation Plugin
PT1	Accelerator pedal	Unused
PT2	Accelerator Pedal	Used
PT3	Brake Pedal	Unused
PT4	Brake Pedal	Used

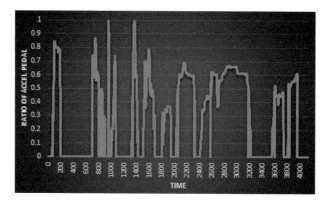

Fig. 3 Driving log of PT1

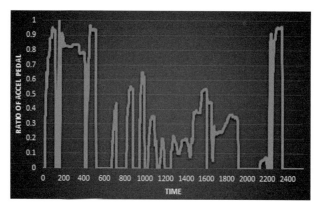

Fig. 4 Driving log of PT2

the plugin is easier than driving without the plugin. Thus, we think that the effect of the navigation plugin occupies a big part of the improvement. In the next section, we classify these original driving logs based on GMM to construct a driver model.

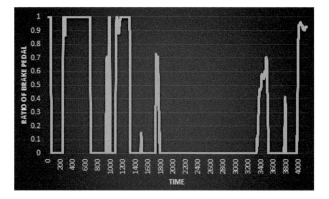

Fig. 5 Driving log of PT3

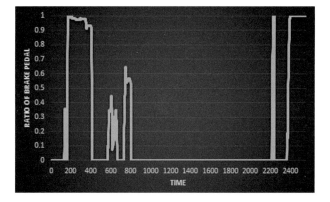

Fig. 6 Driving log of PT4

4 Clustering of Driving Log

Here, we aim for modeling driving behaviors by using the driving log shown in
the previous section. A number of related studies for modeling driving behaviors
have been addressed in the past. Kobayashi et al. proposed an automatic generation
system of driving model from time-series data of vehicles [4]. The time-series data
of vehicles is extracted from movies by a video camera, and the driving model is
based on "Neural Network". The system was able to simulate the behaviors of lead
vehicles and following vehicles. Ueshima et al. also proposed a modeling method
of driver behaviors by using "Bayesian Network" [3]. They considered the change
in brake pedals, but the value of the brake pedals is discrete (i.e., press down or
not). Their method can identify the situation where a driver does not press the brake
pedal. In this paper, we adopted "Gaussian Mixture Model (GMM)" for modeling

driving behaviors according to the method by Mima et al. [9]. Mima et al. considered the time-series data of brake pedal only, on the other hand, we deal with accelerator pedal in addition to brake pedal. Furthermore, we compare driving behaviors between without navigation and with navigation.

GMM is one of the mixture distributions, and GMM consists of some different normal probability distributions. The structure of the GMM is simple, but the approximation capability of the distributions is very high. Moreover, GMM also can be used as an unsupervised clustering algorithm by maximizing a posteriori probability. The probability density function of GMM which consists of M elements is defined by following equation.

$$p(\boldsymbol{x}; \theta) = \sum_{m=1}^{M} \pi_m \mathcal{N}(\boldsymbol{x}; \mu_m, \sigma_m) \tag{1}$$

An element $\mathcal{N}(\boldsymbol{x}; \mu_m, \sigma_m)$ is a normal probability distribution (mean is μ_m and variance is σ_m). Additionally, a parameter π_m should be satisfied following conditions.

$$\pi_m \leq 0 (m = 1, \ldots, M) \tag{2}$$

$$\sum_{m=1}^{M} \pi_m = 1 \tag{3}$$

A parameter π_m represents a selection probability of each element. Thus, a joint probability density that element m outputs x is defined by following equation.

$$p(\boldsymbol{x}, m; \theta) = \pi_m \mathcal{N}(\boldsymbol{x}; \mu_m, \sigma_m) \tag{4}$$

GMM can be regarded as a marginal distribution of the joint probability density. Therefore, a posterior probability that value x is generated from element m is can be calculated by following equation.

$$p(m|\boldsymbol{x}; \theta) = \frac{p(\boldsymbol{x}, m; \theta)}{\sum_{m=1}^{M} \pi_m \mathcal{N}(\boldsymbol{x}; \mu_m, \sigma_m)} \tag{5}$$

The maximization of the posterior probability leads to appropriate clustering results. Consequently, each element (i.e., a normal probability distribution) represents a cluster of $\boldsymbol{x}$.

Here, we classify the driving logs into some clusters by using above method. First of all, we convert the original driving logs to 2-dimensional data. Table 3 shows a partial data of the driving log. The value of "rate" means the pressure rate of the pedal (i.e., the original driving log). The value of "integral" means the integral value

Table 3 Conversion from driving log

Time	Rate	Integral	Time	Rate	Integral
131	0.000	0.000	145	0.215	3.549
132	0.058	0.058	146	0.000	0.000
133	0.074	0.133	147	0.000	0.000
134	0.082	0.215	148	0.000	0.000
135	0.133	0.349	149	0.000	0.000
136	0.152	0.501	150	0.227	0.227
137	0.188	0.690	151	0.631	0.858
138	0.305	0.996	152	1.000	1.858
139	0.407	1.403	153	1.000	2.858
140	0.447	1.850	154	0.843	3.701
141	0.450	2.301	155	0.647	4.349
142	0.415	2.717	156	0.294	4.643
143	0.317	3.035	157	0.000	0.000
144	0.298	3.333	158	0.000	0.000

of the "rate" during the pedal is pressing down. Referring to the Table 3, a driver starts to press the pedal at time 132, and releases the pedal at time 145. The "rate" is accumulated during the time, and the "integral" becomes 3.549 at time 145. This conversion enables to recognize driver's behaviors in non-continuous space.

The estimation of appropriate element model is based on Bayesian Information Criterion (BIC). The line graph of BIC for PT1 is shown in Fig. 7. Each line shows a different type of variance for normal probability distribution. For example "EII" represents the variance of spherical and equal volume. The detail of these descriptions are explained by the official website of Mclust (http://www.stat.washington.edu/mclust/). Mclust is a software package for R (http://www.r-project.org/). The graph indicates that "EVI (diagonal and equal volume)" is the best model for PT1. In addition, the graph also indicates that eight clusters are the best number of elements. However, eight clusters are too many to classify driver's behaviors (Mima et al. adopted four clusters in [9]) and there is little difference of BIC between them, thus we fixed five clusters as an appropriate model in all cases (i.e., PT1, PT2, PT3, and PT4).

The clustering results of all cases are shown in Figs. 8, 9, 10, and 11. The horizontal axis shows the pressure rate of the pedal, and the vertical axis shows the integral of the pressure rate during the pedal is pressing down. We can see that each point belongs to one of the five clusters. A center position of each cluster is summarized in Table 4, and the number of points in each cluster is summarized in Table 5.

Fig. 7 Estimation of
optimal model

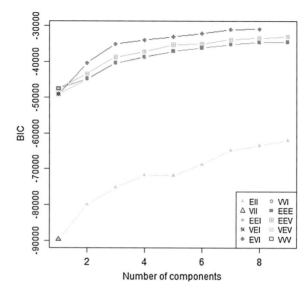

Fig. 8 Clustering result of
PT1

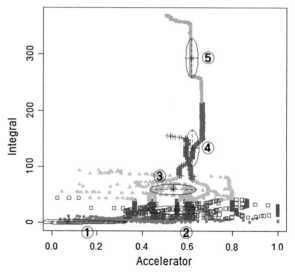

At first, we compare PT1 with PT2. It seems that the clusters of PT1 are arranged
lengthways, on the other hand, the clusters of PT2 are extended breadthways. This
tendency implies that the driver usually keeps the pressure of accelerator pedal (the
pressure rate is approximately from 0.4 to 0.8), but the driver can use both strong
and weak pressures by the influence of navigation facility. In particular, clusters 1
and 2 represent the pressing down weakly (the pressure rate is approximately from
0 to 0.5), and clusters 3, 4, and 5 represent the pressing down strongly (the pressure
rate is approximately from from 0.6 to 0.9). Moreover, Table 5 indicates that most

Fig. 9 Clustering result of
PT2

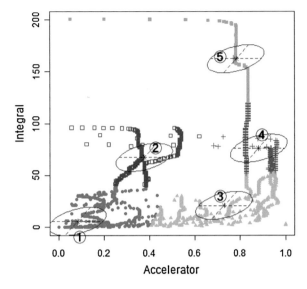

Fig. 10 Clustering result of
PT3

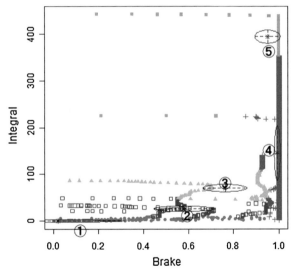

of points belong to cluster 1 (i.e., pressing down the pedal weak and short) in both
patterns, and there is no significant difference among other clusters.

Next, we compare PT3 with PT4. The both patterns indicate that there is a ver-
tically long clusters (4 and 5) at right in the figures (the pressure rate is 1). These
clusters implies the stop of vehicle in front of a traffic light, the pressure of the brake
pedal gradually weakens when the traffic light changes. The most characteristic tendency between the two patterns is the positions of clusters 2 and 3. It seems that
these clusters represent the pressing down weakly to regulate the speed of vehicles.

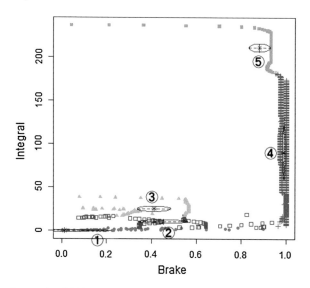

Fig. 11 Clustering result of PT4

Table 4 Center of clusters

		1	2	3	4	5
PT1	Accelerator	0.068	0.548	0.536	0.619	0.617
	Integral	0.843	22.652	59.588	133.341	293.274
PT2	Accelerator	0.081	0.377	0.728	0.881	0.781
	Integral	5.833	67.422	20.952	76.097	162.400
PT3	Brake	0.019	0.577	0.761	0.992	0.950
	Integral	0.160	26.394	70.513	145.234	394.317
PT4	Brake	0.010	0.452	0.411	0.990	0.880
	Integral	0.050	10.634	25.278	90.404	210.847

Table 5 Number of points in clusters

	1	2	3	4	5
PT1	2341 (56.5%)	654 (15.8%)	556 (13.4%)	339 (8.2%)	250 (6.0%)
PT2	1566 (61.3%)	302 (11.8%)	411 (16.1%)	173 (6.8%)	103 (4.0%)
PT3	3059 (73.9%)	180 (4.3%)	164 (4.0%)	642 (15.5%)	95 (2.3%)
PT4	1980 (77.5%)	82 (3.2%)	74 (3.0%)	357 (14.0%)	62 (2.4%)

Additionally, these clusters of PT3 lean to the right, but these clusters of PT4 lean to the left. This result indicates that the navigation facility can suppress sudden braking for drivers. Moreover, Table 5 indicates that the most of points belong to cluster 1 as with the above cases, and it is rare to press down strongly the brake pedal compared to the accelerator pedal.

5 State Transition Graph

The clusters in the previous section showed driver's behaviors in non-continuous space. In this section, we analyze the state transition of drivers in chronological order. Figures 12, 13, 14, and 15 show the state transition diagrams of all cases (i.e., PT1, PT2, PT3, and PT4). Moreover, Tables 6, 7, 8, and 9 show the transition tables among the states of drivers. Each state corresponds to the cluster in the previous section. For example, in Fig. 12, the state of driver changes from 1 to 2 at time 93, from 2 to 3 at time 137, from 3 to 4 at 190, from 4 to 3 at 191, from 3 to 1 at 201, and so on.

At first, we compare PT1 with PT2. The state transition from 1 to 2 in PT1 (17 times) and the state transition from 1 to 3 in PT2 (12 times) are the most frequent pattern in both cases, this transition implies the pressing down the pedal strongly to speed up. Additionally, the state transition from 2 to 3 in PT1 (10 times) is the second frequent pattern, and this transition also implies the pressing down the pedal strongly

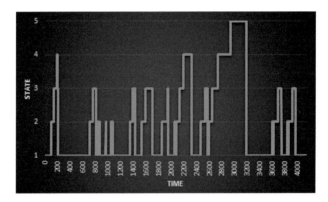

Fig. 12 State transition graph of PT1

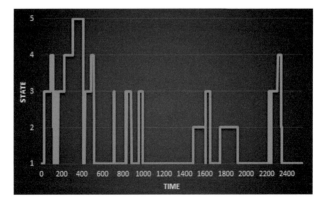

Fig. 13 State transition graph of PT2

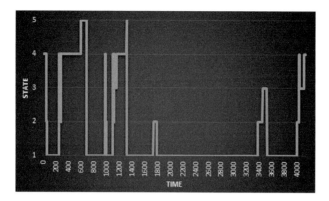

Fig. 14 State transition graph of PT3

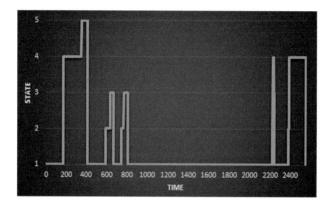

Fig. 15 State transition graph of PT4

Table 6 Transition table of PT1

		To				
		1	2	3	4	5
From	1	0	17	0	0	0
	2	8	0	10	0	0
	3	7	1	0	3	0
	4	1	0	1	0	1
	5	1	0	0	0	0

to speed up as well as the above transition. On the other hand, the state transition from 3 to 1 in PT2 (8 times) is the second frequent pattern, and this transition implies the pressing down the pedal weakly to speed down. It seems that the navigation facility is able to support the speed control for drivers.

Table 7 Transition table of PT2

		To				
		1	2	3	4	5
From	1	0	2	12	0	0
	2	5	0	0	0	0
	3	8	0	0	4	0
	4	0	3	0	0	1
	5	1	0	0	0	0

Table 8 Transition table of PT3

		To				
		1	2	3	4	5
From	1	0	3	0	4	0
	2	3	0	1	4	0
	3	1	1	0	2	0
	4	1	4	3	0	2
	5	2	0	0	0	0

Table 9 Transition table of PT4

		To				
		1	2	3	4	5
From	1	0	5	0	0	0
	2	0	0	3	3	0
	3	3	0	0	0	0
	4	1	1	0	0	1
	5	1	0	0	0	0

Next, we compare PT3 with PT4. In these cases, we focus on the state 3 which implies the keeping up a constant speed. It seems that there are two trends after state 3 in PT3: the pressure rate of brake pedal becomes strong or weak than before. This tendency implies the ambivalence of the driver. On the other hand, there is only one trend after state 3 in PT4: the pressure rate of brake pedal becomes weak than before. This tendency implies that the frequency of sudden brake can be reduced compared to PT3.

Consequently, these results indicate that the novice driver showed different behaviors depending on the navigation facility. Moreover, the control of the speed (i.e., the operation of accelerator and brake pedals) becomes easier by the navigation facility for drivers. However, in this paper, we could only pick up one test driver. Thus, this discussion is not sufficient and limited. We need more driver's logs in order to build a generalized driver model to provide assurance of safety.

6 Conclusion

In this paper, in order to support novice drivers in an unfamiliar driving situation, we developed a virtual driving course based on an actual roundabout intersection located in Ichinomiya, Aichi. Moreover, we analyzed the driving log of a novice driver by using GMM and state transition graphs to evaluate the effect of navigation facility. The results of analysis indicated that the navigation facility we developed as a plugin is able to support novice drivers, and the operations of accelerator and brake pedals become more smoothly. We believe that the knowledge we obtained from these results will be useful to develop more comfortable driving support system for novice drivers.

However, we still have some challenges. A problem we need to take care of first is the generalization of driver's model. We think that the generalized driver model can be used to find traffic traps which novice drivers fall into (e.g., accident-prone area or person). Furthermore, we must cover the possibility of the introduction of new traffic concept called "Shared Space". A city on the concept minimizes the use of traffic lights and traffic signs. The experimental results performed in Kyoto by Miyagawa et al. (http://www.jcomm.or.jp/) indicated that the concept made walking easier without the change of amount of traffic, but the concept has various problems to be solved. Especially, we think that the support for novice drivers by using information technology is a key to success of shared space.

Acknowledgments This work was supported by Grant-in-Aid for Young Scientists (B).

References

1. Abdelhameed, W.A.: Micro-simulation function to display textual data in virtual reality. Int. J. Archit. Comput. **10**(2), 205–218 (2012)
2. Ishikawa, H., Miyajima, C., Kitaoka, N., Takeda, K.: Driving data collection using in-vehicle network and analysis of driving behavior on different types of vehicles. IEICE Tech. Rep. **111**(442), 257–262 (2012)
3. Kamijima, Y., Takei, K., Zuo, Y., Wakita, Y., Kita, E.: Modeling of driver behavior by Bayesian network. Proc. Comput. Mech. Conf. **2011**, 628–629 (2011)
4. Kobayashi, D., Itakura, N., Honda, N., Yikai, K., Kazama, H.: Method of generating a driving model automatically from time series data of vehicles. IPSJ SIG Tech. Rep. **2003**(114), 29–34 (2003)
5. Kok, D., Knowles, M., Morris, A.: Building a driving simulator as an electric vehicle hardware development tool. In: Proceedings of Driving Simulation Conference Europe (2012)
6. Kuroyanagi, Y., Nagasa, Y., Numayama, T., Yamamoto, O., Yamasaki, H., Yamada, M., Yamamoto, S., Nakano, T.: Older driver support: Screening method of cognitive function decline by driving performances. In: Proceedings of ITE Winter Annual Convention, pp. 6–6–1 (2010)
7. Kurumida, K., Kuroyanagi, Y., Miyajima, C., Kitaoka, N., Takeda, K.: A driving diagnosis and instruction system based on past driving data. IEICE Tech. Rep. **110**(381), 87–92 (2011)

8. Matsumura, Y., Tamida, K., Nihei, M., Shino, M., Kamata, M.: Proposal of education method for elderly drivers based on the analysis of their real driving behavior. Proc. TRANSLOG **2011**, 321–324 (2011)
9. Mima, H., Ikeda, K., Shibata, T., Fukaya, N., Hitomi, K., Bando, T.: Estimation of driving state by modeling brake pressure signals. IEICE Tech. Rep. **109**(124), 49–53 (2009)
10. Mirokuji, S., Aso, T., Hasegawa, T.: Performance comparisons between roundabouts and signalized intersections. IEICE Tech. Rep. **109**(459), 113–118 (2010)
11. Nakamura, H., Ryu, K., Sezaki, K., Iwai, M.: Analysis of driver's behavior using mobile phone and vehicle sensors. Proc. Forum Inf. Tech. **12**, 431–432 (2013)
12. Scott, H., Knowles, M., Morris, A., Kok, D.: The role of a driving simulator in driver training to improve fuel economy. In: Proceedings of Driving Simulation Conference Europe (2012)

Bin-Based Estimation of the Amount of Effort for Embedded Software Development Projects with Support Vector Machines

Kazunori Iwata, Elad Liebman, Peter Stone, Toyoshiro Nakashima, Yoshiyuki Anan and Naohiro Ishii

Abstract In this paper we study a bin-based estimation method of the amount of effort associated with code development. We investigate the following 3 variants to define the bins: (1) the same amount of data in a bin (SVM same #), (2) the same range for each bin (SVM same range) and (3) the bins made by Ward's method (SVM Ward). We carry out evaluation experiments to compare the accuracy of the proposed SVM models with that of the ε-SVR using Welch's t-test and effect sizes. These results indicate that the methods SVM same # (1) and SVM Ward (3) can improve the accuracy of estimating the amount of effort in terms of the mean percentage of predictions that fall within 25 % of the actual value.

K. Iwata (✉)
Department of Business Administration, Aichi University,
4-60-6, Hiraike-cho, Nakamura-ku, Nagoya, Aichi 453-8777, Japan
e-mail: kazunori@vega.aichi-u.ac.jp

K. Iwata · E. Liebman · P. Stone
Department of Computer Science, The University of Texas at Austin,
2317 Speedway, Stop D9500, Austin, TX 78712-1757, USA

E. Liebman
e-mail: eladlieb@cs.utexas.edu

P. Stone
e-mail: pstone@cs.utexas.edu

T. Nakashima
Department of Culture-Information Studies, Sugiyama Jogakuen University,
17-3 Moto-machi, Hoshigaoka, Chikusa-ku, Nagoya, Aichi 464-8662, Japan
e-mail: nakasima@sugiyama-u.ac.jp

Y. Anan
Base Division, Omron Software Co., Ltd., Higashiiru, Shiokoji-Horikawa,
Shimogyo-ku, Kyoto 600-8234, Japan
e-mail: yoshiyuki_anan@oss-g.omron.co.jp

N. Ishii
Department of Information Science, Aichi Institute of Technology,
1247 Yachigusa, Yakusa-cho, Toyota, Aichi 470-0392, Japan
e-mail: ishii@aitech.ac.jp

© Springer International Publishing Switzerland 2016
R. Lee (ed.), *Computer and Information Science 2015*,
Studies in Computational Intelligence 614, DOI 10.1007/978-3-319-23467-0_11

1 Introduction

Growth and expansion of the information-based society has resulted in increased use of a wide variety of information products using embedded software systems. The functionality of such products is becoming ever more complex [8, 14], and because of the focus on reliability, guaranteeing product quality is particularly important. Such software represents an important fraction of the budget of businesses and government. It is, therefore, increasingly important for embedded software development companies to realize efficient development methods while guaranteeing delivery time and product quality, and maintaining low development costs [3, 13, 15, 16, 22, 23, 25]. Estimating the amount of effort (man-days cost) requirements for new software projects and guaranteeing product quality are especially important because the amount of effort is directly related to cost, while product quality affects the reputation of the corporation. Considerable attention has been given to various development, management, testing, and reuse techniques, as well as real-time operating systems, tools, and other elements in the embedded software field. However, there has been little research on the relationship between the scale of the development, the amount of effort, and the number of errors using data accumulated from past projects [12, 17, 18]. Thus far, to study the task of effort prediction, the well-known NASA software project data-set has been used [2, 18].

In our formulation of the problem, rather than treat the task of predicting effort as a regression task and predicting a continuous value of effort for code samples, we instead identify blocks of effort, which we refer to as bins, and treat these as labels, which we try to predict, thus treating the problem as a classification task (predicting the correct effort bin for a code sample). In previous work, we investigated the estimation of total effort and errors using artificial neural networks (ANN), and showed that ANN models are superior to regression analysis models for predicting effort and errors in new projects [9, 10]. We also proposed a method to estimate intervals of the number of errors using a support vector machine (SVM) and ANNs [11].

However, these models used a naive method to create bins, which have the same range. In this paper, we propose a novel bin-based estimation method for the amount of effort for embedded software development projects with SVMs, and investigate 3 methods for bin identification. This is crucial to our general framework, since in order to predict an appropriate interval of the amount of effort in a project, it is important to correctly define the intervals (i.e. prediction labels).

In addition, the effectiveness of the SVM (and SVR) using the function depends on selection of the kernel parameter (γ) and soft margin parameter (C) [5]. ε is important for ε-SVR to estimate values effectively. We use three dimensional grid search to select the best combination of them.

We perform extensive evaluations to compare the accuracy of the proposed SVM models with that of the ε-SVR [17] using 10-fold cross-validation as well as by means of Welch's t-test [21, 26] and effect sizes [4, 7]. Our results show that the proposed models can improve the accuracy of estimating the amount of effort in terms of the mean percentage of predictions that fall within 25 % of the actual value.

2 Related Work

2.1 Support Vector Regression

One of the prominent algorithms that has been employed to predict development effort associated with software projects is ε-Support Vector Regression (SVR) [17]. The Support Vector Regression algorithm (SVR) uses the same principles as the canonical Support Vector Machine for classification with a few minor differences [19]. One prominent variant, ε-Support Vector Regression (ε-SVR), uses an ε-insensitive loss function to solve the regression problem and find a closest fitting curve [20].

ε-SVR tries to find a continuous function such that the maximum number of data points lie within the ε-wide insensitivity tube. While previous work did use this approach, it did not probe the optimization of parameters which are crucial to the performance of ε-SVR and similar algorithms, as we do in this paper in Sect. 3.4.

The proposed method to optimize parameters improves the mean magnitude of relative error (*MMRE*: Eq. (3)) from 0.165 [5] to 0.149 by leave-one-out cross-validation (LOOCV) [18]. On the other hand, our proposed SVM models in this paper for the data indicate 0.226 as *MMRE*, because of a small number of data points and independent variables. The number of data points is 18 and that of independent variables is 2.

2.2 Artificial Neural Networks

In earlier papers, we showed that ANN models are superior to regression analysis models for predicting effort and errors in new projects [9]. In addition, we proposed a method for reducing this margin of error [10]. However, methods using ANNs have reached the limit in their improvement, because these methods estimate an appropriate value using what is known as point estimation in statistics. Therefore, we propose in this paper a method for reducing prediction errors using bin-based estimation provided by SVMs. The results of comparison using an ANN are shown in Sect. 4.3. We find out the number of optimal hidden node by 10-fold cross-validation in the comparison. The results demonstrate that the proposed method can estimate the amount of effort better than ANNs.

2.3 Our Contribution

The algorithms proposed in previous work tend to estimate the amount of effort
accurately. However, we maintain that this is to some extent an illusion—the NASA
software project data set includes the small number of data points, and the dispersion
in depended and independent variables is not large. In a more sophisticated approach
like the one we propose, a small data set makes it difficult to create appropriate
bins: performing regression is easier than bin-based estimation in the case of low
dispersion. Our target data sets, however, are large, and manifest a high extent of
variability. Specifically, the amount of effort (the dependent variable) is within a
certain range, but the values of independent variables are highly variable. In this case,
it is difficult for a regression approach to estimate the amount of effort accurately.
Therefore, we propose an approach for creating some kind of bins for projects of
which the amount of effort is within a certain range to reduce the influence of such
dispersion in independent variables.

3 Bin-Based Estimation Models for the Amount of Effort

3.1 Original Data Sets

Using the following data from a large software company, we created bin-based estimation models to estimate the amount of planning effort (Eff).

Eff: "The amount of effort", which indicates man-days cost in a review process for
software development projects.

V_{new}: "Volume of newly added", which denotes the number of steps in the newly
generated functions of the target project.

V_{modify}: "Volume of modification" denoting the number of steps modified or added
to existing functions to use the target project.

V_{survey}: "Volume of original project", which denotes the original number of steps in
the modified functions, and the number of steps deleted from the functions.

V_{reuse}: "Volume of reuse" denoting the number of steps in functions of which only
an external method has been confirmed and which are applied to the target project
design without confirming the internal contents.

3.2 Data Selection for Creating Models

To estimate an appropriate binning for the amount of effort in a project, it is important
to eliminate outliers. Figures 1 and 2 show the distributions of the amount of effort
with bin intervals of 500 and 10, respectively. These distributions confirm that data

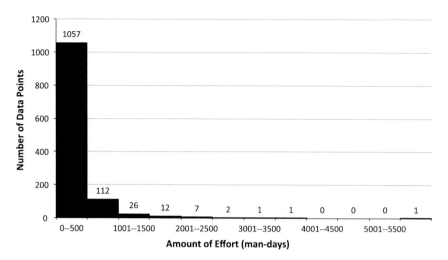

Fig. 1 Distribution of the amount of effort (bins interval 500)

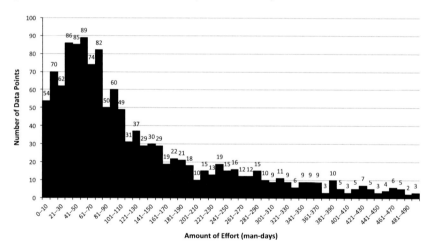

Fig. 2 Distribution of the amount of effort (bins interval 10)

points with less than 500 man-days of effort account for approximately 86.7 % of the total amount of effort. Considering the conditions outlined above, we use the data points which have less than 500 man-days of effort. The distribution of the amount of effort with a bin interval of 10 is shown in Fig. 2. The histogram in this figure has 50 bins and 1057 projects, and our models estimate an appropriate bin for each project.

3.3 General Architecture

SVMs [5, 6] are also supervised learning models. They construct a hyperplane or set of hyperplanes in a high or infinite dimensional space for classification. A good classification can be achieved by the hyperplane with the largest distance to the closest training data point of any class. It often happens, however, that the discrimination sets are not linearly separable in a finite dimensional space. Hence, the SVM maps the original finite dimensional space into a much higher dimensional space in which separation is easier by defining them in terms of a kernel function selected to suit the problem. We use a radial basis function as the kernel function, because this is a popular kernel function for use in SVMs. The corresponding feature space using the function is a Hilbert space of infinite dimensions. Moreover, the effectiveness of the SVM using the function depends on selection of the kernel parameter (γ) and soft margin parameter (C) [5].

The reason why we use SVMs instead of SVRs is that a method to estimate intervals of the number of errors using a support vector machine (SVM) and ANNs showed the better results than these of ANNs for regression and regression analysis [11].

3.3.1 Grouping into Bins for SVM

A representative value of a bin is used as the estimated amount of effort in a project. Therefore, to estimate an appropriate bin of the amount of effort in a project, it is important to define the clusters. We create the following 3 types of bins. A representative value of a cluster is the median of the bin.

- The same amount of data in a bin (SVM same #).
- The same range for each bin (SVM same range).
- The bins made by Ward's method [24] (SVM Ward).

Figure 3 shows the example of same # and same range bins. The target data to be grouped is 15, 20, 30, 40, 50, 70, 80, 90 and 100. The amount of data in each bin is three in the same #. The data belong to the first bin are 15, 20 and 30. The same

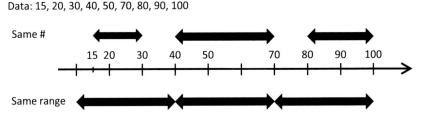

Fig. 3 Example of bins

range adopt 29 as the range. The first bin is [11, 40] and includes 15, 20, 30, 40. If a representative value is the median of each bin, these of the same # are 20, 50 and 90. Correspondingly, these of the same range are 35, 60 and 90.

The accuracy of the estimation depends on the number of bins. Hence, we select the best number of bins with cross-validation and 3D grid-search shown in Sect. 3.4.

3.4 Parameter Selection Using Cross-Validation and 3D Grid-Search

The performance of SVM depends on the choice of the regularization parameters γ and C. The best combination of γ and C is often selected by a grid search with exponentially increasing sequences thereof. In addition, we search for the best number of bins or the most appropriate ε. Hence, we have to define a three-dimensional grid to adapt them using grid-search. The ε and the number of bins are selected with linearly increasing sequences in the three-dimensional grid-search. Figure 4 shows an example of the three-dimensional grid-search. Firstly, the parameters are searched for in the search space $g_1, g_2, \ldots, g_7, g_8$ according to the sparse grid. The cuboid $g'_1, g'_2, \ldots, g'_7, g'_8$ indicating the best combination is found. Next, the cuboid is used as the new search space and partitioned into new grids. Typically, each distinct combination of parameters is checked using cross-validation to avoid over-fitting. We perform 10-fold cross-validation to find the best combination.

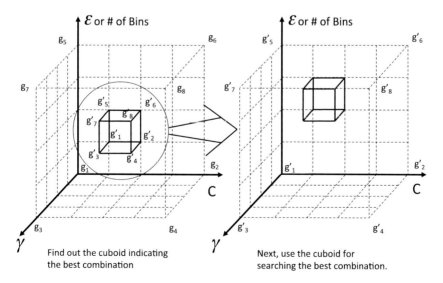

Fig. 4 Example of 3D grid-search

4 Evaluation Experiment

4.1 Evaluation Criteria

The following 6 criteria are used as the performance measures for the effort estimation models [18]. Equations (1) and (3) are, the smaller the value of each evaluation criterion is, the higher is the accuracy. On the other hand, the larger the value of *MPRED*(25) is, the higher is the relative accuracy. The value of $\frac{\widehat{X}-X}{X}$ is regarded as 1, if X is equal to 0 in the calculation of *MARE* and *SDRE*. The accuracy value is expressed as X, while the representative value in the estimated bin is expressed as $\widehat{X}$. A representative value is the median of the bin in this paper. Therefore, if a model could estimate appropriate bins for all projects, *MAE* and *MMRE* would not be 0. For example, if the accuracy value is 13 and the estimated bin is (11, 20], $\widehat{X}$ is 15.5 ((11 + 20)/2) and *MAE* and *MMRE* are equal to 2.5 and 0.1613, respectively. The amount of data is expressed as n.

1. Mean of absolute errors (*MAE*).
2. Standard deviation of absolute errors (*SDAE*).
3. Mean magnitude of relative errors (*MMRE*).
4. Standard deviation of relative errors (*SDRE*).
5. *MPRED*(25) is the mean percentage of predictions that fall within 25 % of the actual value.
6. *SDPRED*(25) is the standard deviation of predictions that fall within 25 % of the actual value.

$$MAE = \frac{1}{n}\sum|\widehat{X} - X| \tag{1}$$

$$SDAE = \sqrt{\frac{1}{n-1}\sum\left(|\widehat{X} - X| - MAE\right)^2} \tag{2}$$

$$MMRE = \frac{1}{n}\sum\left|\frac{\widehat{X} - X}{X}\right| \tag{3}$$

$$SDRE = \sqrt{\frac{1}{n-1}\sum\left(\left|\frac{\widehat{X} - X}{X}\right| - MARE\right)^2} \tag{4}$$

4.2 Data Used in Evaluation Experiment

We performed 10-fold cross validation on data from 1057 real projects in the evaluation experiment. The original data were randomly partitioned into 10 equal sized subsamples (with each subsample having data from 105 or 106 projects). One of the subsamples was used as the validation data for testing the model, while the remaining nine subsamples were used as training data. The cross-validation process was repeated ten times with each of the ten subsamples used exactly once as validation data.

4.3 Results and Discussion

For each model, the experimental results of the 10-fold cross validation are shown in Tables 1, 2 and 3.

We compared the accuracy of the proposed models with that of the ε-SVR using Welch's t-test [26] and effect sizes [4, 7]. A Student's t-test [21] is used to test the null hypothesis that the means of two normally distributed populations are equal. Welch's t-test is used when the variances of the two samples are assumed to be different to test the null hypothesis that the means of two normally distributed populations are equal if the two sample sizes are equal [1]. Given the t-value and degrees of freedom, a p-value can be found using a table of values from the Student's t-distribution. If the p-value is smaller than or equal to the significance level, the null hypothesis is rejected. The null hypothesis in our experiment is interpreted as "there is no difference between the means of the estimation errors (or the mean percentage) for the proposed model and ε-SVR". Effect size measures either the sizes of associations or the sizes of differences. Cohen provided rules of thumb for interpreting these effect sizes,

Table 1 Experimental results (absolute errors) for estimating the amount of effort

	MAE	SDAE	95 % Confidence interval
SVM same #	37.546	38.437	[35.226, 39.866]
SVM same range	40.568	41.689	[38.052, 43.084]
SVM ward	38.311	40.384	[35.874, 40.748]
ε-SVR	36.669	39.403	[34.291, 39.047]
ANN model	84.169	60.449	[80.521, 87.817]

Table 2 Experimental results (relative errors) for estimating the amount of effort

	MMRE	SDRE	95 % Confidence interval
SVM same #	0.65355	1.0157	[0.59225, 0.71485]
SVM same range	0.74389	1.3956	[0.65966, 0.82812]
SVM ward	0.68157	1.1862	[0.60998, 0.75316]
ε-SVR	0.71025	2.0037	[0.58932, 0.83118]
ANN model	0.96687	0.082109	[0.96191, 0.97183]

Table 3 Experimental results (PRED(25)) for estimating the amount of effort

	MPRED(25)	SDPRED(25)	95% Confidence interval
SVM same #	0.36558	0.05924	[0.32320, 0.40796]
SVM same range	0.31064	0.03924	[0.28257, 0.33871]
SVM Ward	0.35707	0.04098	[0.32775, 0.38639]
ε-SVR	0.30305	0.04505	[0.27082, 0.33528]
ANN model	0.0038005	0.000024074	[0.0037833, 0.0038177]

suggesting that Cohen's d of $|0.1|$ represents a 'small' effect size, $|0.3|$ represents a 'medium' effect size and $|0.5|$ represents a 'large' effect size.

The results of the t-test and Cohen's d for *MAE*, *MMRE* and *MPRED(25)* in estimating the amount of effort are given in Tables 4, 5 and 6. The underlined p-values in the tables indicates statistically significant differences between the type of bin and ε-SVR. In addition, the underlined Cohen's d values in the tables mean the effect size is large.

Tables 1 and 4 indicate that the method of SVM same range cannot improve the accuracy to estimate the amount of effort than that of ε-SVR in *MAE* and the others have the same estimating accuracy as ε-SVR. The Tables 2 and 5 mean that the proposed methods have the same estimating accuracy as ε-SVR in *MMRE*. The results for *MPRED(25)* indicate that statistically significant differences between SVM same # and ε-SVR, and SVM Ward and ε-SVR. In addition, SVM same # and SVM ward improve about 6.252 % $(= \sqrt{(0.05924^2 + 0.04505^2)/2} \times 1.188)$ and 5.400 % $(= \sqrt{(0.04098^2 + 0.04505^2)/2} \times 1.254)$ in terms of *MPRED(25)*, respectively.

Table 4 Results of t-test for *MAE* between each type of bin and ε-SVR

	SVM same #	SVM same range	SVM ward
t-value	0.5180	2.210	0.9462
p-value	0.6045	**0.02723**	0.3422
Cohen's d	0.02253	0.09612	0.04115

Table 5 Results of t-test for *MMRE* between each type of bin and ε-SVR

	SVM same #	SVM same range	SVM ward
t-value	0.8206	0.4479	0.4004
p-value	0.4210	0.6543	0.6889
Cohen's d	0.03569	0.01948	0.01741

Table 6 Results of t-test for $MPRED(25)$ between each type of bin and ε-SVR

	SVM same #	SVM same range	SVM Ward
t-value	3.082	0.4017	2.805
p-value	**0.006835**	0.6927	**0.01178**
Cohen's d	**1.188**	0.1797	**1.254**

It is evident from these results that the methods SVM same # and SVM Ward can improve the accuracy of estimating the amount of effort in terms of the mean percentage of predictions that fall within 25 % of the actual value. However, the methods and SVM same range cannot improve the mean of absolute errors and the mean magnitude of relative errors. The cause of the results is several large errors for estimating in proposed methods. Despite the usefulness of the mean to investigate the accuracy of models, outliers have the biggest effect on the mean.

5 Conclusion

In this paper we have discussed a bin-based estimation method for the amount of effort with SVMs and investigated the following three approaches for defining suitable bins: (1) the same amount of data in a bin (SVM same #), (2) the same range for each bin (SVM same range) and (3) the bins made by Ward's method (SVM Ward). We have carried out evaluation experiments to compare the accuracy of the proposed SVM model with that of the ε-SVR using 10-fold cross-validation as well as by means of Welch's t-test and effect sizes. The results in estimating the amount of effort have indicated statistically significant differences between SVM same # and ε-SVR, and SVM Ward and ε-SVR in terms of $MPRED(25)$. In addition, SVM same # and SVM ward have improved $MPRED(25)$ about 6.252 % and 5.400 %, respectively. These results have exhibited that the methods SVM same # and SVM Ward can improve the accuracy of estimating the amount of effort in terms of the mean percentage of predictions that fall within 25 % of the actual value.

Our future research includes the following:

1. Having implemented a model to estimate the final amount of effort in new projects, we plan to estimate the amount of effort at various stages in the project development process (e.g. halfway).
2. We intend to employ a more complex method to improve the overall prediction accuracy.
3. Since outliers can be detrimental to our model, more refined approaches to outlier detection may be beneficial to our framework.
4. Overall, more data is needed to further support our work.

Acknowledgments A portion of this work has taken place in the Learning Agents Research Group (LARG) at the Artificial Intelligence Laboratory, The University of Texas at Austin. LARG research is supported in part by grants from the National Science Foundation (CNS-1330072, CNS-1305287), ONR (21C184-01), AFRL (FA8750-14-1-0070), and AFOSR (FA9550-14-1-0087).

References

1. Aoki, S.: In testing whether the means of two populations are different (in Japanese) (2007). http://aoki2.si.gunma-u.ac.jp/lecture/BF/index.html
2. Bailey, J.W., Basili, V.R.: A meta-model for software development resource expenditures. In: Proceedings of the 5th International Conference on Software Engineering, ICSE'81, pp. 107–116. IEEE Press, Piscataway (1981). http://dl.acm.org/citation.cfm?id=800078.802522
3. Boehm, B.: Software engineering. IEEE Trans. Softw. Eng. **C-25**(12), 1226–1241 (1976)
4. Cohen, J.: Statistical Power Analysis for the Behavioral Sciences, 2nd edn. Routledge, New York (1988). http://www.worldcat.org/isbn/0805802835
5. Cortes, C., Vapnik, V.: Support-vector networks. Mach. Learn. **20**(3), 273–297 (1995)
6. Cristianini, N., Shawe-Taylor, J.: An Introduction to Support Vector Machines and Other Kernel-Based Learning Methods. Cambridge University Press, Cambridge (2000)
7. Cumming, G.: The new statistics: why and how. Psychol. Sci. **25**(1), 7–29 (2014)
8. Hirayama, M.: Current state of embedded software (in Japanese). J. Inf. Process. Soc. Jpn. (IPSJ) **45**(7), 677–681 (2004)
9. Iwata, K., Nakashima, T., Anan, Y., Ishii, N.: Error estimation models integrating previous models and using artificial neural networks for embedded software development projects. In: Proceedings of 20th IEEE International Conference on Tools with Artificial Intelligence, pp. 371–378 (2008)
10. Iwata, K., Nakashima, T., Anan, Y., Ishii, N.: Improving accuracy of an artificial neural network model to predict effort and errors in embedded software development projects. In: Lee, R., Ma, J., Bacon, L., Du, W., Petridis M. (eds.) Software Engineering, Artificial Intelligence, Networking and Parallel/Distributed Computing. Studies in Computational Intelligence, vol. 295, pp. 11–21. Springer, Heidelberg (2010). doi:10.1007/978-3-642-13265-0_2
11. Iwata, K., Nakashima, T., Anan, Y., Ishii, N.: Estimating interval of the number of errors for embedded software development projects. Int. J. Softw. Innov. (IJSI) **2**(3), 40–50 (2014). doi:10.4018/ijsi.2014070104
12. Kemerer, C.F.: An empirical validation of software cost estimation models. Commun. ACM **30**(5), 416–429 (1987). doi:10.1145/22899.22906
13. Komiyama, T.: Development of foundation for effective and efficient software process improvement (in Japanese). J. Inf. Process. Soc. Jpn. (IPSJ) **44**(4), 341–347 (2003)
14. Nakamoto, Y., Takada, H., Tamaru, K.: Current state and trend in embedded systems (in Japanese). J. Inf. Process. Soc. Jpn. (IPSJ) **38**(10), 871–878 (1997)
15. Nakashima, S.: Introduction to model-checking of embedded software (in Japanese). J. Inf. Process. Soc. Jpn. (IPSJ) **45**(7), 690–693 (2004)
16. Ogasawara, H., Kojima, S.: Process improvement activities that put importance on stay power (in japanese). J. Inf. Process. Soc. Jpn. (IPSJ) **44**(4), 334–340 (2003)
17. Oliveira, A.L.: Estimation of software project effort with support vector regression. Neurocomputing **69**(1315), 1749–1753 (2006). doi:10.1016/j.neucom.2005.12.119. http://www.sciencedirect.com/science/article/pii/S0925231205004492
18. Shin, M., Goel, A.: Empirical data modeling in software engineering using radial basis functions. IEEE Trans. Softw. Eng. **26**(6), 567–576 (2000). doi:10.1109/32.852743
19. Smola, A., Scholköpf, B.: A tutorial on support vector regression. Stat. Comput. **14**(3), 199–222 (2004). doi:10.1023/B:STCO.0000035301.49549.88

20. Smola, A.J., Schölkopf, B.: A tutorial on support vector regression. Stat. Comput. **14**(3), 199–222 (2004). doi:10.1023/B:STCO.0000035301.49549.88
21. Student: The probable error of a mean. Biometrika **6**(1), 1–25 (1908)
22. Takagi, Y.: A case study of the success factor in large-scale software system development project (in Japanese). J. Inf. Process. Soc. Jpn. (IPSJ) **44**(4), 348–356 (2003)
23. Tamaru, K.: Trends in software development platform for embedded systems (in Japanese). J. Inf. Process. Soc. Jpn. (IPSJ) **45**(7), 699–703 (2004)
24. Ward, J.H.: Hierarchical grouping to optimize an objective function. J. Am. Stat. Assoc. **58**(301), 236–244 (1963). doi:10.1080/01621459.1963.10500845
25. Watanabe, H.: Product line technology for software development (in Japanese). J. Inf. Process. Soc. Jpn. (IPSJ) **45**(7), 694–698 (2004)
26. Welch, B.L.: The generalization of student's problem when several different population variances are involved. Biometrika **34**(28), 28 (1947)

Applying RoBuSt Method for Robustness Testing of the Non-interference Property

Maha Naceur and Lilia Sfaxi

Abstract When setting up a secure system, rigorous testing is important to implement and sustain a system that will induce customer confidence. In order to improve the testing process of security properties, formal methods of specification are developed to automatically generate tests. In this work, we propose to apply an approach we developed in a previous work to test the robustness of a very restrictive and important security property, which is non-interference. We consider the case of distributed component-based systems, where avoiding interference can represent a real challenge, especially when exchanging messages between heterogeneous entities.

1 Introduction

When setting up a secure system, rigorous testing is important to implement and sustain a system that will induce customer confidence. Tests detect the presence of errors in a system operation. There are various types of tests that should be performed, including security testing, software and hardware reliability, and compatibility between all the elements of the system. Testing is an important validation activity. This activity is based on different techniques, the best known are: static analysis, model-checking and testing as a genuine validation activity. Testing is a difficult, expensive, time-consuming and labor-intensive process and it should be repeated each time an implementation, referred to as an Implementation Under Test (IUT), is modified. A promising improvement of the testing process is to automatically generate tests from formal models of specification. Robustness testing is a part of the validation process. It allows us to check if IUT still fulfills some robustness requirements. In general, the robustness of a secure system is its ability to respond

M. Naceur (✉)
LIP 2 Laboratory, University of Tunis El Manar, Tunis, Tunisia
e-mail: naceur.maha@gmail.com

L. Sfaxi (✉)
LIP 2 Laboratory, University of Tunis El Manar, INSAT, University of Carthage,
Tunis, Tunisia
e-mail: liliasfaxi@gmail.com

© Springer International Publishing Switzerland 2016
R. Lee (ed.), *Computer and Information Science 2015*,
Studies in Computational Intelligence 614, DOI 10.1007/978-3-319-23467-0_12

correctly against exceptional or unforeseen execution conditions such as the unavailability of system resources, communication failures, invalid or stressful inputs, etc.

To automate robustness testing, most of approaches are based on fault-injection techniques which are particularly adequate to test large software systems. Several tools implement those techniques such as FUZZ [8], BALLISTA [4], JCrasher [3]…. Other approaches are based on abstract specification of the system behavior in order to select test inputs [10]. We also cite the PROTOS project [5] that consists in mutating the system behavior to introduce abnormal inputs and tests are generated by performing simulations on this specification.

As we are interested in black-box testing, we were inspired from works presented in [11] which presents principles and techniques for model based black-box conformance testing of real-time systems using the Uppaal model-checking tool-suite [17]. The basis for testing is given as a network of concurrent timed automata specified by the tester. It outlines two extreme approaches to timed test generation: Offline and Online test generation. In this paper, we use the online testing approach (on-the-fly). This approach combines test generation and execution. The test generator interactively interprets the model, and observes the IUT behavior. Only a single test input is generated from the model at a time, then immediately executed on the IUT. The output produced by the IUT is checked against the specification, new inputs are produced until the tester decides to end the test, or an error is detected.

Most of the classical security mechanisms for insuring the confidentiality and integrity properties in a system are based on access control and cryptography. These mechanisms are efficient when restricting access to the data to a defined number or quality of users at a given place and time, but fail to control the propagation of information across the system. That's why information flow control policies are natural for specifying end-to-end confidentiality and integrity requirements. The information flow policy we adopt in this paper is *non-interference*. This property requires that confidential data must not affect the publicly visible behavior of the system [1]. This definition implies that a mechanism is needed to follow all information flows that circulate in the system, whether explicitly or implicitly, then to check if this information exchange respects the initial constraints of the system. *Non-interference* is a very restrictive security property, and real systems, most of the times, need to intentionally release some private information. So we need another security property, coupled to non-interference, that checks whether any release of information, intentional or not, can affect the security of the information flow. This property is *robustness*.

Non-interference robustness has already been addressed in several works, such as [15]. But these works are dedicated to enforcing a semantic security property of robustness, and address especially decentralized systems (that imply the presence of a set of attackers, each of them having the power to act on a set of security configurations). In our case, *robustness testing* is targeted, which consists on validating each system before its deployment. We also consider mostly centralized non-interference, where the security configuration is handled by a single actor, that represents a potential attacker, each time he decides to modify this configuration. To the best of our knowledge, no previous works have addressed robustness testing of non-interference.

This article is composed of four main sections. The next section presents the robustness method we use to automate tests generation. The second section presents the notion of non-interference and how it is applied to component-based systems. In the third section, we briefly explain how we represent the concepts above with our formal model. The last section, before the conclusion and perspectives, shows an illustrative use case.

2 The Robustness Method

In this work, we adopt a robustness testing approach called **RoBuSt** presented in [16]. The main goal of this approach is to validate security protocols in the case of a stressful behavior of the environment in which the system evolves. The robustness test method allows to generate tests from the specification of the system's behavior in a hostile environment. Then, the tester runs these test cases to check if the system is robust or not. The steps of this method are illustrated in the Fig. 1.

This method is presented as follows:

(1) Initially, we express the nominal specification of the system under test, which describes the nominal behavior of that system in the nominal conditions of the environment. In the case of security protocol, nominal conditions represent the absence of an attack.

(2) Suspension traces (the useful actions that should be insured by the IUT) are added to the nominal specification to obtain an increased specification, which represents the behavior of the system in a hostile environment. The system must continue to work correctly if an attack is produced.

(3) We determine in the next step the Robustness Test Purpose (RTP), describing the behavior of the system that the tester intends to observe. This phase allows to

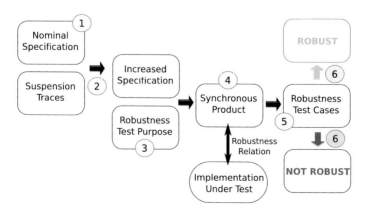

Fig. 1 Robustness testing architecture

derive and select the execution traces assimilating invalid and/or inappropriate entries and acceptable exits. An RTP corresponds to a property or a need that the tester wishes to observe in the implementation under test. Under the robustness test, an RTP describes some aspects of operations in the hostile environment. In our approach, RTP is defined as a reachability property which asks whether a given state formula j can possibly be satisfied by any reachable location. In other words: "Is there a path starting at the initial state such as j is eventually satisfied along that path?".

(4) When synchronizing the increased specification of a system under test and the RTP, a synchronous product is obtained. It contains all the execution sequences of the IUT.

(5) We generate all execution sequences called Robustness Test Cases (RTC), a set of conditions under which a tester will determine whether a system is working correctly or not. RTCs are generated from the synchronous product between the increased specification and the RTP.

(6) An RTC announces if the system is "Robust" or "NotRobust" against unexpected attacks.

3 Non-Interference

3.1 Non-interference and Security Levels

Non Interference is an information flow control model initially defined by Goguen and Meseguer [9]. Its main goal is to warrant that sensitive data do not affect the public behavior of the system. This security property is insured by the control of information flows in the system, and the end-to-end application of confidentiality and integrity properties.

Let's take for example an explicit information flow between two variables. Let's consider the code variables l and h, with l representing a public integer (with a low security level) and h a secret integer (with a high security level). The simple assignment $(l := h + 3)$ is illegal, because the secret value contained in h will flow to l, a public variable. In this case, any modification of h will induce a modification of l, and thus enable any random user to guess the value of h. On the other hand, the instruction: $(h := l - 3)$ is legal, because the public variable does not depend on the secret variable.

In the real world, security levels are not as simple as *public* and *private*. For example, a root password to a server is supposed to be secret to most of the employees, but not to the system administrator and to the owner. The main issues in this case will be: how to define the different levels of security in a system, and how to attach them to data? For this purpose, **security labels** are used. A label attached to an object specifies how to use the information in this object, by assigning a security level. The set of all security labels in a system represent a **security lattice**. To insure that an information transmission is non-interferent, we have to make sure that it flows to a more (or equally) restrictive target, from the confidentiality and integrity point of view. Thereby, we are sure that any secret or sensitive information is taken into consideration, not divulged to a party that is not allowed to handle it.

In the literature, many label models are defined, such as [12–14]. The model we use in this work is inspired by the model defined in [13], that represents labels as a set of tags. A tag is an opaque term that, taken individually, does not have any significant meaning, but when attached to data, assigns to it a certain level of confidentiality or integrity. A label l belonging to a set of security levels L, contains a couple of sets: S (for Secrecy), representing the confidentiality level, and I (for Integrity), representing the integrity level. Thereby, $l = \{S; I\}$.

Confidentiality Assigning a confidentiality tag j ($j \in S$) to an object o means that o contains a private information with a level of confidentiality j. Any other object in the system *receiving* an information flow from o must have at least the confidentiality tag j.

Integrity The integrity level i ($i \in I$) of an object represents a guarantee of the authenticity of the information in this object. It gives an idea to the receiver of the object about which parties have tainted (modified) the information. Assigning i to an object o represents an additional warranty, so the more integrity tags are attached to o, the less reliable it is. Therefore, any object in the system that *transmits* an information flow to o must have at least the integrity tag i.

Partial order relation The set of labels in a systems is governed by a partial order relation *can flow to*, represented by the symbol $\subseteq$. This relation orders labels in the lattice, from the least restrictive to the most restrictive. It gathers two subset relations: $\subseteq_C$ and $\subseteq_I$

Fig. 2 Example of a security lattice

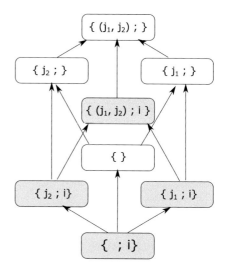

- Having two labels $l_1 = \{S_1; I_1\}$ and $l_2 = \{S_2; I_2\}$, $l_1 \subseteq l_2$ if and only if $S_1 \subseteq_C S_2$ and $I_1 \subseteq_I I_2$
- $S_1 \subseteq_C S_2$ if and only if $\forall j_1 \in S_1, \exists j_2 \in S_2$ where $j_1 = j_2$
- $I_1 \subseteq_I I_2$ if and only if $\forall i_2 \in I_2, \exists i_1 \in I_1$ where $i_1 = i_2$

The Fig. 2 shows an example of a security lattice, using security tags.

If two labels are not connected in the lattice, and no path allows to go from one to the other (as for $\{j1; \}$ and $\{j2; i\}$), they are considered incomparable, and noted: $l_1 \nsubseteq l_2$ and $l_2 \nsubseteq l_1$.

3.2 Non-interference in Component-Based Distributed Systems

Following the information flow in a centralized system can be quite tedious, and is even more so in a distributed system. Among the existing solutions that consider non-interference in distributed systems, some of them, as [6, 19], consider information flow in a rather coarse granularity level, which may allow certain information leakage due to implicit flows, and others, like [7, 18], although considering fine grained information flow control, base the system distribution on security constraints rather than on functional needs.

To avoid these drawbacks, [1] use the component-based model to represent distributed systems. A component is a composition unit that can be deployed independently and assembled with other components. Thanks to their modularity and loose coupling, the development and management of distributed systems are greatly simplified. Each component contains several configuration interfaces (called attributes), server and client ports for message exchange, and bindings defining its connections with other components.

To check whether component-based systems are non-interferent, [1] suggest affecting labels only to ports and attributes of a component, as shown in Fig. 3. Information flows will then be considered from two points of view: intra-component and inter-component.

Fig. 3 Assigning labels to a component-based system

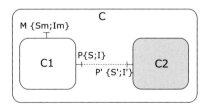

- Intra-component security is checked for each component whose code is available, by applying a tool called CIFIntra on its implementation. It will check whether interferences exist in the code of the component.
- Inter-component security is checked for every binding in the system: a binding is authorized between a port P1 and a port P2 if and only if $label(P1) \subseteq label(P1)$.

This method is interesting in many ways:

- The security labels are assigned at a high level, while the non-interference is checked at a very fine grain.
- Separating the functional behavior and the security behavior enables the separation of concerns, and allows multiple actors to continue their work without being obliged to interfere with the area of expertise of one another. By actors, we mean functional architect, security architect, developer and tester.
- The component model helps separating the internal behavior of the functional units, and their interaction, thanks to the components' interfaces. In this way, we can proceed with testing the intra-component and inter-component behaviors separately.

In fact, in our work, we are interested in testing the robustness of the **inter-component** non-interference property of a given component-based distributed system. The system representation is based on the model described in [1], but, unlike them, we consider not only the static non-interference (labels assignment and verification at compile-time), but also dynamic non-interference: if the security architect decides to change the security configuration at runtime, how will the system react? Will a statically non-interferent system remain non-interferent?

3.3 Non-interference and Downgrading

The non-interference property is very restrictive. In fact, it is impossible to find a real system entirely non-interferent. For example, a simple password verification process represents an interference, as the value of public output of the system (the server's response, whether the password is correct or not) depends on the private input (the password). In some cases, like this one, an interference is considered mandatory, or having a minor effect on the global security of the system. That's why, a mechanism must exist, showing which interferences are authorized. This mechanism is called "downgrading" (more specifically, *declassification* for confidentiality and *endorsement* for integrity). Downgrading a label means decreasing its security level, making it less restrictive. In the case of confidentiality, declassification is done by removing tags, since every tag adds more restrictiveness. Dually, endorsing an integrity label means adding new integrity tags.

4 Robustness Testing of Non-interferent Component-Based Distributed Systems

4.1 Main Issues

Robustness testing enables to check whether an implementation under test (IUT) can function correctly in the presence of invalid inputs or stressful environmental conditions. Robustness testing is done by defining a set of controlled inputs that may be either valid or invalid which allow the tester to know exactly what to expect. In the case of an initially non-interferent component-based system, we need to ensure that the system is still non-interferent under invalid entries. So, assigning or modifying the label of a communication port can be considered dangerous, if it produces an interference, that is, a *forbidden* interference.

We distinguish several types of attacks: passive or active. On the one hand, passive attackers are able to observe the state of a computational system as it evolves. On the other hand, active attackers are able not only to observe the behavior of the system but also to modify it. An attack can also be internal or external to the system. In our case, we start by studying the dynamic behavior of a component-based system if faced with an *active internal* attack. If we consider a distributed system that is initially non-interferent, many events that occur during its execution can induce information flow leaks: the replacement of a faulty component, the addition or removal of a binding, the modification of the security configuration…. In our case study, we consider this latter (modification of the security configuration), more precisely the modification of a component's security label, as being the "invalid input" to be tested. Furthermore, we consider the case of a centralized security model, which means that the security configuration is defined and controlled by a central entity, the *Security Architect*.

To summarize, the attack we are considering is *internal*, because the attacker is the Security Architect, a legal actor in the system; and *active*, as this architect will change some of the security labels, whether intentionally or unintentionally.

4.2 Concepts Modeling

We introduce the concepts and notations used throughout the paper. We used Uppaal [17], a real-time systems' validation tool. It is designed in order to verify system operations such as protocols or multimedia applications modeled as Network of Timed Automata (NTA).

4.2.1 Component-Based System

A component-based system (CBS) is described as an NTA. CBS is composed of a set of components. Each component is modeled as a timed automata (TA) [2]. In the following, we use X to denote the set of clocks which are incremented when components exchange messages. A TA is a tuple defined as follows:

$$TA = (Q, q_0, X, Act, E, I)$$

where: Q is a finite set of locations in which a component is at runtime, q_0 is the initial location where a component has not yet started the exchange of messages, $Act = Act_{In} \cup Act_{Out} \cup \{\tau\}$ is a finite set of input actions (denoted by *message?*: receiving a message), output actions (denoted by *message!*: sending a message) and un-observable actions ξ, $E \subseteq Q \times Act \times G(X) \times 2^X \times Q$ is the set of transitions between locations with actions, a guard and a set of clocks to be reset and $I : Q \rightarrow G(X)$ assigns invariants to locations if necessary. A state of the NTA is defined by locations of all TAs, the clock values, and discrete variables and denoted as follow: $\bar{q} = (q_1, \ldots, q_n)$. Every TA may fire a transition separately or synchronize with another TA, which leads to a new location. The semantics of an NTA is defined as a labeled transition system $(S_Q, s_0, \hookrightarrow)$, where $S_Q = (Q_1 \times \cdots \times Q_n) \times \mathbb{R}^X$ is the set of states, $s_0 = (\bar{q}_0, u_0)$ is the initial state and $\hookrightarrow \subseteq S \times S$ is the transition relation.

4.2.2 Security Configuration

To assign the security configuration to a component-based system, we use discrete variables. To do this, five matrices are defined: **BIND** (*port* $\times$ *port*), where a cell (p_1, p_2) indicates if a binding between p_1 and p_2 exists; **Integrity** (*tag* $\times$ *port*), used to assign an integrity label to a port. A label being a set of tags, each cell (t_i, p) of the matrix indicates if the tag t_i exists in the label of p; **Confidentiality** (*tag* $\times$ *port*), where a cell (t_c, p) indicates if the confidentiality tag t_c exists in the confidentiality label of p; **B_A_I** (*binding* $\times$ *tag*), where a cell (b, t_i) indicates if the endorsement of the integrity tag t_i is authorized for the binding b and **B_A_C** (*binding* $\times$ *tag*), where a cell (b, t_c) indicates if the declassification of the confidentiality tag t_c is authorized for the binding b.

4.2.3 The Attack

An *Attack* is also defined as a TA as follows:

$$Attack = (Q^A, q_0^A, X, Act^A, E^A, I^A)$$

An attack is launched after the initialization phase. In our case, every time the architect changes the security labels at runtime, the system checks if this change causes an interference. If it's the case, the binding (or bindings) concerned with the interference is (are) broken. Nevertheless, a broken binding can be of no consequence to the normal execution of the system, if this binding is not used after the attack is triggered. That's why the system will be considered interferent if and only if a component tries to use a broken binding. In that case, every other actor of the system is notified that there is an interference, and the execution is stopped.

5 Application to a Simple Use-Case

Let's take for example two components representing respectively a developer and a tester (illustrated in Fig. 4). The developer writes a code snippet, sends it to the tester (via BIND1), who tests it. If a bug is detected, the tester sends the bug type and description to the developer (via BIND2), that fixes it and sends the code back to the tester for validation (via BIND3). Each time a component sends a message, the clock value is incremented.

5.1 Security Configuration

A security label is assigned to each port of each component. We consider here four tags:

- *dc*: confidentiality tag, assigned to each data that contains a secret of the developer.
- *tc*: confidentiality tag, assigned to each data that contains a secret of the tester.
- *di*: integrity tag, assigned to each data that can be modified by the developer.
- *ti*: integrity tag, assigned to each data that can be modified by the tester.

According to these tags and to the definition of our label model defined in Sect. 3.1, a data that has both the confidentiality tags *dc* and *tc* is more restrictive than the one having only one of the tags, because it contains both secrets. Dually, a data having integrity tags *di* and *ti* is less restrictive that the one having only ti, because it can be altered by more actors, thus being less reliable.

The Fig. 5 shows the resulting matrices: BIND, Integrity and Confidentiality.

Initially, the security configuration is defined as follows:

- All the exchanged messages have the confidentiality level dc,tc, as the codes and the test results can be viewed by both the developer and the tester.

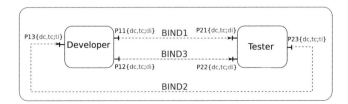

Fig. 4 A simple component based system

BIND	P11	P12	P13	P21	P22	P23
P11	0	0	0	1	0	0
P12	0	0	0	0	1	0
P13	0	0	0	0	0	1
P21	1	0	0	0	0	0
P22	0	1	0	0	0	0
P23	0	0	1	0	0	0

CONF	dc	tc
P11	1	1
P12	1	1
P13	1	1
P21	1	1
P22	1	1
P23	1	1

INTEG	di	ti
P11	1	0
P12	1	0
P13	0	1
P21	1	0
P22	1	0
P23	0	1

Fig. 5 BIND, confidentiality and integrity matrices

Fig. 6 B_A_I and B_A_C matrices

B_A_C	dc	tc
BIND1	0	0
BIND2	0	0
BIND3	0	0

B_A_I	di	ti
BIND1	0	0
BIND2	0	0
BIND3	0	0

- The codes sent to the tester can only be altered by the developer (hence the *di* tag).
- The test results (bugs) can only be altered by the tester (*ti* tag).

On the other hand, for our first test, we consider that the matrices B_A_I and B_A_C are filled with zeros, as represented in the Fig. 6. Indeed, we consider that, at first, no downgrading is authorized.

5.2 Nominal Specification of the Use-Case

The nominal specification of the IUT is modeled by an NTA (Fig. 7), composed of three TAs:

- *Nominal_Architect* = $(Q^{NA}, q_0^{NA}, X, Act^{NA}, E^{NA}, I^{NA})$: represents the behavior of the security architect, where Q^{NA} = {INIT, BEGIN}, q_0^{NA} = {INIT}, Act^{NA} = {∅}, E^{NA} = {(INIT, Initialization(), BEGIN)} meanning that the architect reaches the BEGIN location after initializing the configuration matrices and I^{NA} = {∅}.
- *Nominal_Developer* = $(Q^{ND}, q_0^{ND}, X, Act^{ND}, E^{ND}, I^{ND})$: represents the component Developer in its nominal behavior, where Q^{ND} = {INIT, SEND, RECEIVE}, q_0^{ND} = {INIT}, Act^{ND} = {BIND1!, BIND2?, BIND3!}, E^{ND} = {(INIT, BIND1!, SEND), (SEND, BIND2?, RECEIVE), (RECEIVE, BIND3!, INIT)} and I^{ND} = {∅}.

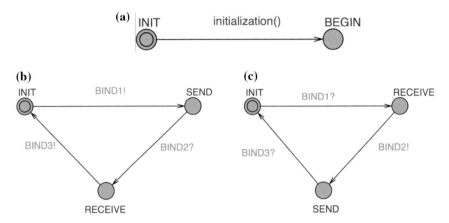

Fig. 7 Nominal specification. **a** Nominal architect, **b** Nominal developer, **c** Nominal tester

- *Nominal_Tester* $= (Q^{NT}, q_0^{NT}, X, Act^{NT}, E^{NT}, I^{NT})$: represents the Tester behavior, and defines the same locations as the developer, but with dual actions, where $Q^{NT} = \{\text{INIT, RECEIVE, SEND}\}$, $q_0^{NT} = \{\text{INIT}\}$, $Act^{NT} = \{\text{BIND1?, BIND2!,}$ $\text{BIND3?}\}$, $E^{NT} = \{(\text{INIT, BIND1?, RECEIVE}), (\text{RECEIVE, BIND2!, SEND}),$ $(\text{SEND, BIND3?, INIT})\}$ and $I^{NT} = \{\emptyset\}$.

It is defined as follows:

$$NTA_{NS} = (Q^{NS}, q_0^{NS}, X, Act^{NS}, E^{NS}, I^{NS})$$

where $Q^{NS} = Q^{NA} \cup Q^{ND} \cup Q^{NT}$ is the set of vectors, $q_0^{NS} = (q_0^{NA}, q_0^{ND}, q_0^{NT})$ is the initial vector, $Act^{NS} = Act^{NA} \cup Act^{ND} \cup Act^{NT}$ the set of actions of CBS and $E^{NS} = E^{NA} \cup E^{ND} \cup E^{NT}$.

5.3 Increased Specification

A robustness requirement aims at ensuring that the system will always be non-interferent under non nominal execution conditions. So, we increase the nominal specification by adding suspension traces. The first step done by the architect is to check the initial security configuration for any interferences, before authorizing the execution. If there is no attack, or if the attack is authorized, the behavior of the network is almost the same as the one in nominal conditions. The only difference is that each component has to check if a binding is valid before crossing it. In our system, we add three locations to the nominal specification of the architect (Fig. 8), which are:

- **CHECK**, where the architect checks if there is an interference,
- **ATTACK**, where the architect simulates an attack,
- **INTERFERENCE**, to which the system evolves if an interference is identified.

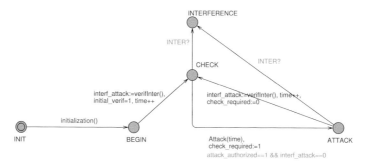

Fig. 8 Increased architect

The TA corresponding to the increased architect becomes:

$$Increased\_Architect = (Q^{IA}, q_0^{IA}, X, Act^{IA}, E^{IA}, I^{IA})$$

It represents the behavior of the active internal attack causing by the architect, where Q^{IA} = {INIT, BEGIN, CHECK, ATTACK, INTERFERENCE}, Act_{IA} = {INTER?}, $q_0^{IA} = q_0^{NA}$ and E^{IA} = {(INIT, Initialization(), BEGIN), (BEGIN, VerifInter(), CHECK), (CHECK, Attack(Time), ATTACK), (ATTACK, VerifInter(), CHECK), (ATTACK / CHECK, INTER?, INTERFERENCE)}.

We also add suspension traces to the specification of components. A location "Interference" is added: if this component tries to cross a broken binding, it moves to this location and broadcasts an alert (INTER) to the other actors of the system. The latters, when receiving the alert, move to their "Interference" locations. Figure 9 shows the TA corresponding to the Increased Developer in a hostile environment.

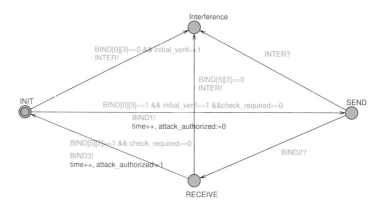

Fig. 9 Increased developer

The TA corresponding to the increased Developer becomes:

$$Increased\_Developer = (Q^{ID}, q_0^{ID}, X, Act^{ID}, E^{ID}, I^{ID})$$

It represents the behavior of the Developer in a hostile environment, where $Q^{ID} =$ {INIT, SEND, RECEIVE, Interference}, $Act_{ID} = $ {INTER!, INTER?}, $q_0^{ID} = q_0^{ND}$ and $E^{ID} = \{E^{ID} \cup$ (INIT/RECEIVE, INTER!, Interference) $\cup$ (SEND, INTER?, Interference) }.

The TA corresponding to the increased Tester becomes:

$$Increased\_Testesr = (Q^{IT}, q_0^{IT}, X, Act^{IT}, E^{IT}, I^{IT})$$

It represents the behavior of the Tester in a hostile environment, where $Q^{IT} = $ {INIT, RECEIVE, SEND, Interference}, $Act_{IT} = $ {INTER!, INTER?}, $q_0^{IT} = q_0^{NT}$ and $E^{IT} = \{E^{IT} \cup$ (INIT/SEND, INTER!, Interference) $\cup$ (RECEIVE, INTER?, Interference) }.

All TAs, assembled together, represent the increased specification. When an attack is produced, the architect creates two other matrices: Integrity Attack Matrix $IA(port \times tag)$ and Confidentiality Attack Matrix $CA(port \times tag)$, where it specifies the new values of the labels. An attack is produced by the architect at a certain time, while crossing the transition leading to the location ATTACK. We define the increased specification using the following formula:

$$NTA_{IS} = Increased\_Architect \cup Increased\_Developer \cup Increaseed\_Tester$$

5.4 Synchronous Product

An RTP is intended to check if the system is behaving in a correct way. In our example, it helps to observe the consequences of an attack generated at a given time, and check whether it leads to an interference. A set of reachability properties is defined, as for example the property : **E<> interf_attack==i**, that means: "Is there a path from the initial state where a given attack i always causes an interference?".

Once the adequate reachability properties chosen, we model the RTP using a TA with three locations: *INIT*, *Accept* and *Reject*. The *Reject* location means that the RTP is not accepted by the system. In this case, all attacks creating an interference lead the RTP to the state *Reject*. Otherwise, the RTP reaches the state *Accept*. It is modeled as follows:

$$RTP = (Q^{RTP}, q^{0RTP}, X, Act^{RTP}, E^{RTP}, I^{RTP})$$

Where $Q^{RTP} = $ {INIT, Accept, Reject}, $q_0^{RTP} = $ {INIT}, $Act^{RTP} = $ {INTER?}, $E^{RTP} =$ {(INIT, *Interf_attack* == 1, Reject), (INIT, *Interf_attack* == 0, Accept)} and $I^{RTP} = \emptyset$. Figure 10 represents the TA corresponding to a robustness test purpose.

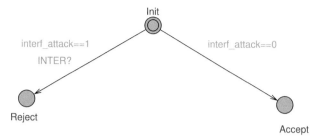

Fig. 10 Robustness test purpose

Intuitively, a synchronous product is obtained by combining the increased specification and the RTP. It contains all the execution sequences that may be produced
either if there is an attack or not.

A synchronous product is defined by *Syn* such as:

$$Syn = NTA_{IS} \otimes RTP$$

Where $Q_{Syn} = \{(q1, q2) \mid q1 \in Q_{IS}, q2 \in Q_{RTP}\}$, $q_0^{Syn} = (q_0^{IS}, q_0^{RTP})$, $X_{Syn} = X$,
$Act_{Syn} = Act_{IS} \cup Act_{RTP}$, $E_{Syn} = E_{IS} \cup E_{RTP}$ and $I_{Syn} = \emptyset$.

5.5 Generation and Execution of Robustness Test Cases from the Synchronous Product

Robustness test cases (RTCs) are a set of conditions under which a tester will determine whether a system is working correctly or not. RTCs are generated from the
synchronous product. An RTC announces if the system is "**Robust**" or "**NotRobust**"
against unexpected attacks. It is defined as an execution sequence which is a finite
path in the labeled transition system of the synchronous product with an observation
clock that records the global elapsed time since the beginning of the computation.
Formally, RTC is defined as follows:

$$RTC = (s_0, 0) \hookrightarrow (s_1, \tau_1) \hookrightarrow \ldots \hookrightarrow (s_n, \tau_n),$$

where $s_i = (\overline{q}_i, v_i)$ and $\tau_{i-1} \leq \tau_i$ for $i = 1 \ldots n$ where s is a state in the synchronous
product and τ is a time value. With Uppaal [17], an RTC is a message Sequence
Chart (MSC) generated when starting the simulator. MSC represents an execution
sequence, starting from the initial location of the synchronous product and ending
by a deadlock location where the RTP stops in one of the two locations *Accept* or
Reject.

In our use case, we suppose that, while the system is running, the security architect
chooses to allow the component "tester" to modify the code he receives from the

Fig. 11 Non-robust test case

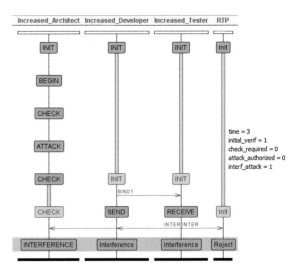

component "developer", in order to perform some unit tests. To do so, he adds the integrity tag *ti* to the port P21, through which the component "tester" receives the code from the "developer". This modification represents a runtime attack, and the system tries to check if this attack can be harmful. This is the case when the destination port of the binding becomes more restrictive than the source port, and if an endorsement is not allowed for this port.

This RTC is represented by the MSC of the Fig. 11.

5.6 Verdicts

Finally, we extract RTCs that satisfy the RTP. The test case execution delivers a verdict that is either "*the system is robust*" (for example: RTC stops at INIT locations for all components) or "*the system is not robust*" (for example: RTC reaches the Interference location for all components).

In the case of the non-interference property, a Not Robust verdict is obtained if the attack causes an interferent information flow. For each binding, we check if the attack causes an interference, i.e. if the label of the client port is not less restrictive than that of the server port. If it's the case, it checks if this interference is an authorized one, i.e. if downgrading the tag causing the interference is authorized for the binding in question, by consulting the matrices B_A_I and B_A_C. If it's not, the binding is broken, and the attack number and time are stored. However, the components are not aware of the attacks the architect is generating. They are affected by an attack only if it generates a forbidden interference that breaks a binding they are about to cross. This is why, not only the label change is important, but also its occurrence time: the same attack can be innocuous if produced after the targeted binding is crossed.

In our example, represented by the Fig. 11, the verdict is that the system is not robust. In fact, the message, sent from the port P11 (with an integrity tag di) flows to the port P21 (with the modified integrity tag di,ti), which is less restrictive. In this case, an interference is detected. The system then checks the matrix B_A_I to see if the endorsement of the tag ti is allowed for the binding BIND1, which is not. This binding is then broken, and the system moves to the "interference" state the next time the developer tries to send a code to the tester via BIND1. If we suppose that the value of the cell (BIND1,ti) in the matrix B_A_I is "1", the system considers that the endorsement of the tag ti for the binding BIND1 is allowed, and the input is considered valid.

6 Conclusion

Robustness testing can be a very efficient way to detect the interference problems that can occur at compile or run-time, even before the deployment of the system. In order to test the robustness of the non-interference property for component-based systems, we applied RoBuSt, an approach that allows us to automatically generate robustness test cases. The obtained set of robustness test cases checks whether this type of systems is robust regarding to dynamic security level changes or not. When executing several tests on our running example, we noticed that a supposedly non-interferent attack can become interferent when coupled with other attacks. Indeed, the verdict of robustness testing does not only depend on the action performed by an attack, but also on the time of its occurrence, and the order of the preceding attacks.

In this work, we considered only inter-component interference checking. However, interferences happen mainly in the implementation of each component. That's why our next target is to test the robustness of component-based systems at both the inter-component and intra-component level. On the other hand, robustness testing can help us be even more expressive: instead of determining if a label change is interferent or not, it would be more realistic to determine how harmful an interference can be for the whole system, and instead of having only two verdicts (robust or not-robust), we should have several shades of robustness levels.

References

1. Abdellatif, T., Sfaxi, L., Robbana, R., Lakhnech, Y.: Automating information flow control in component-based distributed systems. In: Proceedings of the 14th International ACM Sigsoft Symposium on Component Based Software Engineering, CBSE'11, pp. 73–82. ACM, New York (2011)
2. Alur, R., Dill, D.L.: A theory of timed automata. Theor. Comput. Sci. **126**(2), 183–235 (1994)
3. Csallner, C., Smaragdakis, Y.: Check 'n' crash: combining static checking and testing. In: Proceedings of the 27th International Conference on Software Engineering, 2005, pp. 422–431 (2005)

4. Dix, M., Hofmann, H.D.: Automated software robustness testing—static and adaptive test case design methods. In: Proceedings of the 28th International Conference on Euromicro Conference, pp. 62–66 (2002)
5. Du, W., Mathur, A.P.: Vulnerability testing of software system using fault injection. Technical report (1998)
6. Eyers, D.M., Roberts, B., Bacon, J., Papagiannis, I., Migliavacca, M., Pietzuch, P., Shand, B.: Event-processing middleware with information flow control. In: Proceedings of the 10th ACM/IFIP/USENIX International Conference on Middleware, Middleware'09, pp. 32:1–32:2. Springer, New York (2009)
7. Fournet, C., Guernic, G.L., Rezk, T.: A security-preserving compiler for distributed programs: from information-flow policies to cryptographic mechanisms. In: Proceedings of the 16th ACM Conference on Computer and Communications Security, CCS'09. ACM, New York (2009)
8. Fu, Y., Kon, O.: Security and robustness by protocol testing. IEEE Syst. J. **1**, 99 (2012)
9. Goguen, J.A., Meseguer, J.: Security policies and security models. In: Proceedings of IEEE Symposium on Security and Privacy, pp. 11–20 (1982)
10. Helmy, A., Gupta, S.: Fotg: fault-oriented stress testing of ip multicast. IEEE Commun. Lett. **9**(4), 375–377 (2005)
11. Hessel, A., Larsen, K.G., Mikuèionis, M., Nielsen, B., Pettersson, P., Skou, A.: Testing real-time systems using uppaal (2008)
12. Khair, M., Mavridis, I., Pangalos, G.: Design of secure distributed medical database systems. In: Proceedings of the International Conference on Database and Expert systems Applications (1998)
13. Krohn, M.: Information flow control for standard os abstractions. In: Proceedings of Twenty-first ACM SIGOPS Symposium on Operating Systems Principles, SOSP'07, ACM. New York (2007)
14. Myers, A.C., Liskov, B.: Protecting privacy using the decentralized label model. ACM Trans. Softw. Eng. Methodol. (TOSEM) **9**(4), 410–442 (2000)
15. Myers, A.C., Sabelfeld, A., Zdancewic, S.: Enforcing robust declassification and qualified robustness. J. Comput. Secur. **14**(2), 157–196 (2006)
16. Naceur, M., Sfaxi, L., Robbana, R.: Robustness testing for secure wireless sensor network. In: Proceedings of the International Conference on Automation, Control, Engineering and Computer Science, ACECS'14, Monastir, Tunisia (2014)
17. Upsala University. Uppaal tool. www.uppaal.org (2014)
18. Zdancewic, S., Zheng, L., Nystrom, N., Myers, A.C.: Secure program partitioning. ACM Trans. Comput. Syst. **20**(3), 283–328 (2002)
19. Zeldovich, N., Boyd-Wickizer, S., Mazières, D.: Securing distributed systems with information flow control. In: Proceedings of the 5th USENIX Symposium on Networked Systems Design and Implementation, NSDI'08, pp. 293–308. USENIX Association, Berkeley (2008)

An Improved Multi-SOM Algorithm for Determining the Optimal Number of Clusters

Imèn Khanchouch, Malika Charrad and Mohamed Limam

Abstract The interpretation of the quality of clusters and the determination of the optimal number of clusters is still a crucial problem in cluster Analysis. In this paper, we focus in on multi-SOM clustering approach which overcomes the problem of extracting the number of clusters from the SOM map through the use of a clustering validity index. We test the multi-SOM algorithm using real and artificial data sets with different evaluation criteria not used previously such as Davies Bouldin index, and Silhouette index. The multi-SOM algorithm is compared to k-means and Birch methods. Results developed with R language show that it is more efficient than classical clustering methods.

Keywords Clustering · SOM · Multi-SOM · DB index · Dunn index · Silhouette index

1 Introduction

Clustering is considered as one of the most important tasks in data mining. It is a process of grouping similar objects or elements of data set into classes called clusters.

The main idea of clustering is to partition a given set of data points into groups of similar objects where the notion of similarity is defined by a distance function. In the

I. Khanchouch (✉) · M. Limam
Department of Computer Science, University of Tunis, ISG, Tunis, Tunisia
e-mail: imen.khanchouch@yahoo.kr

M. Charrad
Department of Computer Science, RIADI lab, University of Gabès, Gabès, Tunisia
e-mail: malika.charrad@gmail.com

M. Charrad
MSDMA Team, Cedric, CNAM, Paris, France

M. Limam
Department of Statistics, Dhofar University, Salalah, Oman
e-mail: mohamed.limam@isg.rnu.tn

© Springer International Publishing Switzerland 2016
R. Lee (ed.), *Computer and Information Science 2015*,
Studies in Computational Intelligence 614, DOI 10.1007/978-3-319-23467-0_13

literature there are many clustering methods such as hierarchical, partition-based, density-based and neural networks (NN) and each one has its advantages and limits.

We focus on neural networks especially Self Organizing Map (SOM) method. SOM is proposed by [1], it is the most widely used neural network method based on an unsupervised learning technique.

SOM method aims to reduce a high dimensional data to a low dimensional grid by mapping similar data elements together. This grid is used to visualize the whole data set. However, SOM method suffers from the delimitation of clusters, since its main function is to visualize data in the form of a map and not to return a specified number of clusters. That's why a multi-SOM approach has been proposed by [2] to overcome this limit. To return the optimal number of clusters, [3] integrated a cluster validity index called Dynamic Validity Index (DVI) into the multi-SOM algorithm. Then, it is interesting to test this algorithm with other existing validity criteria.

In this paper, we study the existing clustering evaluation criteria and test multi-SOM with different validity indexes, then compare it with a partitioning and a hierarchical clustering method. We used R as a statistical tool to develop the multi-SOM algorithm. The rest of this paper is structured as follows. Section 2 describes different clustering approaches. Section 3 details the multi-SOM approach and provides a literature review. Clustering evaluation criteria are given in Sect. 4. Finally, a conclusion and some future work are given in Sect. 5.

2 Clustering Methods

2.1 Hierarchical Methods

Hierarchical methods aim to build a hierarchy of clusters with many levels. There are two types of hierarchical clustering approaches namely agglomerative methods (bottom–up) and divisive methods (Top–down).

Divisive methods begin with a sample of data as one cluster and successively divide clusters as objects. However, the clustering in the agglomerative methods start by many data objects taken as clusters and are successively joined two by two until obtaining a single partition containing all objects.

The output of hierarchical methods is a tree structure called a dendrogram which is very large and may include incorrect information. Several hierarchical clustering methods have been proposed such as: CURE [4], BIRCH [5], and CHAMELEON [6].

2.2 Partitioning Methods

Partitioning methods divide the data set into disjoint partitions where each partition represents a cluster. Clusters are formed to optimize an objective partitioning criterion, often called a similarity function, such as distance. Each cluster is represented by a centroid or a representative cluster. Partitioning methods such as K-means [7], and PAM [8], suffer from the sensitivity of initialization. Actually, inappropriate initialization may lead to bad results.

2.3 Density-Based Methods

Density-based clustering methods aim to discover clusters with different shapes. They are based on the assumption that regions with high density constitute clusters, which are separated by regions with low density. They are based on the concept of cloud of points with higher density where the neighborhoods of a point are defined by a threshold of distance or number of nearest neighbors. Several density-based clustering methods have been proposed such as: DBSCAN [9] and OPTICS [10].

2.4 Neural Networks

NN are complex systems with high degree of interconnected neurons. Unlike the hierarchical and partitioning clustering methods NN contains many nodes or artificial neurons so it can accept a large number of high dimensional data. Many neuronal clustering methods exist such as SOM and Neural Gas.

In the training process, the nodes compete to be the most similar to the input vector node. Euclidean distance is commonly used to measure distances between input vectors and output nodes' weights. The node with the minimum distance is the winner, also known as the Best Matching Unit (BMU). The latter is a SOM unit having the closest weight to the current input vector after calculating the Euclidean distance from each existing weight vector to the chosen input record. Therefore, the neighbors of the BMU on the map are determined and adjusted. The main function of SOM is to map the input data from a high dimensional space to a lower dimensional one. It is appropriate for visualization of high-dimensional data allowing a reduction of data and its complexity. However, SOM map is insufficient to define the boundaries of each cluster since there is no clear separation of data items. Thus, extracting partitions from SOM grid is a crucial task. In fact, SOM output does not automatically give partitions, but its major function is to visualize a low dimensional map reduced from a high dimensional input data. Also, SOM initializes the topology and the size of the grid where the choice of the size is very sensitive to the generalization of the method. Hence, we extend multi-SOM to overcome these shortcomings and give the optimal number of clusters without any initialization.

3 Multi-SOM Method

3.1 Definition

The multi-SOM is an unsupervised method introduced by [1]. Its main idea is the superposition and the communication between many SOM maps. It builds a hierarchy of SOM maps as shown in Fig. 1.

  The input data are firstly clustered using the SOM. Then, other levels of data are clustered iteratively based on the first SOM grid. Clusters are extracted from each SOM grid by calculating the DB index.

  We proposed to start with a SOM grid of 6 ∗ 6. The size of the maps decreases gradually since the optimal number of clusters is obtained in the last layer. Each grid gathers similar elements into groups from the previous layer. This approach can overcome the problem of cluster determination of the SOM grid.

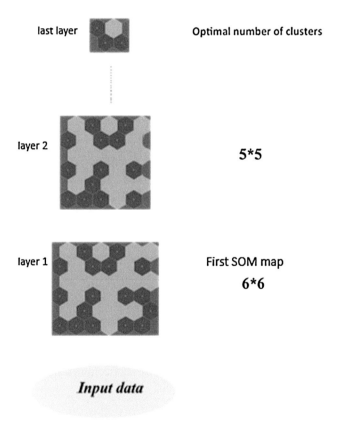

Fig. 1 Multi-SOM architecture

3.2 Literature Review

The Multi-SOM method was firstly introduced by [1] for scientific and technical information analysis specifically for patenting transgenic plant to improve the resistance of the plants to pathogen agents.

Reference [1] proposed an extension of SOM called multi-SOM to introduce the notion of viewpoints into the information analysis with its multiple maps visualization and dynamicity. A viewpoint is defined as a partition of the analyst reasoning.

The objects in a partition could be homogenous or heterogeneous and not necessary similar. However objects in a cluster are similar and homogenous where a criterion of similarity is inevitably used. Each map in multi-SOM represents a viewpoint and the information in each map is represented by nodes (classes) and logical areas (group of classes).

Reference [11] applied multi-SOM on an iconographic database. Iconographic is the collected representation illustrating a subject which can be an image or a document text. Then, multi-SOM model is applied in the domain of patent analysis in [12, 13], where a patent is an official document conferring a right. The experiments use a database of one thousand patents about oil engineering technology and indicate the efficiency of viewpoint oriented analysis, where selected viewpoints correspond to; uses advantages, patentees and titles of patents.

Reference [2] applied multi-SOM algorithm to macrophage gene expression analysis. Their proposed algorithm overcomes some weaknesses of clustering methods which are the cluster number estimation in partitioning methods and the delimitation of partitions from the output grid of SOM algorithm. The idea of [2] consists on obtaining compact and well separated clusters using an evaluation criterion namely DVI. The DVI metric is derived from compactness and separation properties. Thus, compactness and separation are two criteria to evaluate clustering quality and to select the optimal clustering layer.

Reference [14] applied multi-SOM to real data sets to improve multi-SOM algorithm introduced by [2].

3.3 The Proposed Algorithm

Our proposed algorithm aims to find the optimal clusters using the DB index as an evaluation criterion instead of DVI in the algorithm of [3]. We chose to use in this work the DB index because it is similar to DVI since lower values they indicate both, better clustering quality is obtained. DB index is well used in many works and it belongs to the same group of internal criteria which is based on the compactness and separation of the clusters. However, DVI index uses several operations unlike DB index which reduces the memory space and the number of iterations.

First, the whole dataset is introduced as input to the multi-SOM algorithm in the first step to be clustered by SOM and the output is the first SOM grid SOM1 that will be partitioned. We have to mention the size of the first SOM map in advance namely the grid height Hs and the grid width Ws. The multi-SOM algorithm uses the function Batch map proposed by [15] which is a version of SOM Kohonen algorithm; it is faster than SOM in the training. Then, we proceed to the second step which is the clustering of the SOM map so the size of the grid decreases iteratively from a level to another and we compute the DB index at each level and compare therefore different maps until we obtain the optimal number of clusters which is the minimal value of DB index.

The number of neurons decreases iteratively from a grid to another even the DB index. Also we proposed another termination criterion to avoid useless computation and iterations instead of stopping the training when only one neuron is found in the algorithm of [3]. Hence, we stop the training at the level when the DB index takes its minimum value and the optimal number of clusters is found.

The different steps of our algorithm are given as follow:
s: is the SOM layer current number
Hs: the SOMs grid height
Ws: the SOMs grid width
Is: the input of SOMs
max_it: the maximal iterations number for training the SOM grids
Input: W1, H1, I1
Output: Optimal cluster number

```
Begin

• Step1: Clustering data by SOM
s = 1;
Batch SOM (W1, H1, I1, max_it);
Compute DB index;
s = s+1;
• Step2: Clustering of the SOM and cluster delimitation
Hs = Hs - 1;
Ws = Ws - 1;
Repeat
Batch SOM (W1, H1, I1, max_it);
Compute DB index on each SOM grid;
s=s+1;
Until (DBs < DBs+1);
Return (Data partitions, optimal cluster number);
End
```

4 Clustering Evaluation Criteria

The main problem in clustering is to determine the ideal number of clusters. Thus, cluster evaluation is usually used.

In fact, many techniques and measures are used to test the quality of the clusters obtained as output data.

There are three categories of cluster evaluation namely: External validity measures, internal validity measures and relative validity measures.

- External criteria are based on the prior knowledge about data. They measure the similarity between clusters and a partition model. It is equivalent to have a labeled dataset. Many external criteria are cited in the literature like purity, entropy and F-measure.
- Relative criteria are based on the comparison of two different clusters or clustering results. The most known index is: SD index proposed by [16].
- Internal criteria are often based on compactness and separation. That's why in this work we focus on the internal validity indexes to check the quality of clusters.

Compactness is assessed by the intra-distance variability which should be minimized. Separation is assessed by the inter-distance between two clusters which should be maximized.

Many internal criteria exist such as: DB, Dunn, Silhouette, C, CH, DVI, etc. But, we focus on the following indexes:

- *Davies-Bouldin (DB)*

DB is proposed by [17], and given by:

$$DB = \frac{1}{c} \sum_{i=1}^{c} \max_{i \neq j} \left\{ \frac{d(X_i) + d(X_j)}{d(c_i, c_j)} \right\} \qquad (1)$$

where c defines the number of clusters, i and j are the clusters, $d(X_i)$ and $d(X_j)$ are distances between all objects in clusters i and j to their respective cluster centroids, and d (ci, cj) is the distance between these two centroids. Small values of DB index indicate good clustering quality.

- Dunn Index (DI)

DI is proposed by [18], and given by:

$$DI = \min_{1 \leq i \leq c} \left\{ \min \left\{ \frac{d(c_i, c_j)}{\max_{1 \leq k \leq c} (d(X_k))} \right\} \right\} \qquad (2)$$

where d(ci, cj) denotes the distance between c_i and c_j.

$d(X_k)$ represents the intra-cluster distance of the cluster X_k and c is the cluster number of the dataset. Larger values of DI indicate better clustering quality.

- Dynamic Validity Index (DVI)

The DVI metric is introduced by [19], derived from compactness and separation properties. It considers both the intra-distance and the inter-distance which are defined as follows:

$$DVI = \min_{k=1\ldots K} \{IntraRatio\,(k) + \gamma\,InterRatio\,(k)\} \tag{3}$$

$$IntraRatio(l) = \frac{Intra\,(l)}{MaxIntra\,(l)} \tag{4}$$

$$InterRatio(l) = \frac{Inter\,(l)}{MaxInter\,(l)} \tag{5}$$

where l is the layer of each grid and:

$$MaxIntra = \max_{l \in \{1..L\}}(Intra\,(l)) \tag{6}$$

$$Intra\,(l) = \frac{1}{N} \sum_{i=1}^{k_l} \sum_{x \in C_i} (\|x - Z_i\|)^2 \tag{7}$$

$$MaxInter = \max_{l \in \{1..L\}}(Inter\,(l)) \tag{8}$$

$$Inter\,(l) = \frac{\max\left(\|Z_i - Z_j\|^2\right)}{\min_{i \neq j}\left(\|Z_i - Z_j\|^2\right)} \sum_{i=1}^{k_l} \left(\frac{1}{\sum_{j=1}^{kl}\left(\|Z_i - Z_j\|^2\right)}\right)$$

where N is the number of data samples, Z_i and Z_j represent the reference vectors of nodes i and j, γ is a modulating parameter and l denotes a given layer of a SOM grid.

The optimal number of clusters is determined by the minimal value of DVI in each level.

- Silhouette

This measure, introduced by [20], is defined by:

$$S = \frac{b\,(i) - a\,(i)}{\max\{a\,(i),b\,(i)\}} \tag{9}$$

where a (i) is the average distance between the ith sample and all samples included in X_j, b (j) is the minimum average distance between the ith and all samples clustered in X_k ($\square = 1\ldots\square; \square \neq \square$). Larger values of Silhouette index indicate better clustering quality.

5 Experiments

In this section, we carry out the evaluation of the multi-SOM algorithm on artificial and real data sets with different clustering validity indexes as shown in Tables 1 and 2.

5.1 Artificial Data Sets

The artificial data sets used contain 2, 3, 5, or 8 distinct non overlapping clusters. The data sets consist of a total of 300 points each. The clusters are embedded in a 2, 3 or 8 dimensional Euclidean space and have curves, circles, rectangles or ellipses shapes. The actual distribution of the points within clusters followed multivariate normal distribution.

5.2 Real Data Sets

Wine [21] is the real data set used in this paper. It is the result of a chemical analysis of wines derived from 3 different cultivars so this analysis determines the quantities of 13 constituents found in each of the three types of wines which are: Alcohol, Malic acid, Ash, Alcalinity of ash, Magnesium and total phenols.

5.3 Evaluation of Experimental Results

We use $6 * 6$ as the dimension of the SOM map for the first SOM grid and we use R statistical language to test the clustering methods with different validity indices.

The number of classes in Wine data set is equal to 3 which is equal to the obtained number of clusters with multi-SOM method. Then, the obtained number of clusters

Table 1 Evaluation of the multi-som algorithm on Wine data set

Method	The obtained number of clusters	Optimal values of indexes		
		DB	Silhouette	Dunn
Multi-SOM	3	0.4	0.63	0.56
K-means	5	0.49	0.55	0.53
Birch	4	0.51	0.29	0.44

Table 2 Evaluation of the multi-som algorithm on artificial data sets

Method	The obtained number of clusters	DB	Silhouette	Dunn
Circular Datasets	Smpcircles (**C2**)			
Multi-SOM	2	0.41	0.38	0.47
K-means	2	0.58	0.22	0.39
Birch	2	0.69	0.55	0.52
	Smpcircles (**C3**)			
Multi-SOM	3	0.35	0.26	0.66
K-means	2	0.44	0.22	0.47
Birch	2	0.44	0.21	0.41
	Smpcircles (**C5**)			
Multi-SOM	5	0.4	0.36	0.488
K-means	4	0.42	0.32	0.45
Birch	3	0.61	0.28	0.33
	Smpcircles (**C8**)			
Multi-SOM	8	0.33	0.28	0.44
K-means	7	0.51	0.16	0.41
Birch	6	0.53	0.15	0.38
Rectangular Datasets	Smprect (**R2**)			
Multi-SOM	2	0.27	0.25	0.64
K-means	2	0.38	0.53	0.72
Birch	2	0.46	0.61	0.74
	Smprect (**R3**)			
Multi-SOM	3	0.43	0.27	0.44
K-means	2	0.45	0.39	0.48
Birch	2	0.51	0.51	0.56
	Smprect (**R5**)			
Multi-SOM	5	0.47	0.26	0.72
K-means	5	0.61	0.24	0.66
Birch	3	0.58	0.22	0.61
	Smprect (**R8**)			
Multi-SOM	8	0.22	0.34	0.57
K-means	6	0.43	0.27	0.49
Birch	6	0.44	0.26	0.45
Elliptical Datasets	SmEllip (**E2**)			
Multi-SOM	2	0.43	0.26	0.42
K-means	2	0.46	0.25	0.37
Birch	2	0.52	0.2	0.26

(continued)

Table 2 (continued)

Method	The obtained number of clusters	DB	Silhouette	Dunn
SmEllip (**E3**)				
Multi-SOM	**3**	0.34	0.28	0.54
K-means	2	0.45	0.21	0.41
Birch	3	0.47	0.13	0.22
SmEllip (**E5**)				
Multi-SOM	**5**	0.39	0.37	0.49
K-means	4	0.55	0.35	0.45
Birch	3	0.4	0.33	0.39
SmEllip (**E8**)				
Multi-SOM	**8**	0.38	0.21	0.73
K-means	7	0.4	0.12	0.71
Birch	6	0.507	0.11	0.67

by k-means method is 5 and by Birch is 4, so we conclude that the correct number of clusters is obtained by multi-SOM as shown in Table 1.

The value of DB index in Tables 1 and 2 corresponds to the minimal value in different levels of SOM maps. However, the value of silhouette and Dunn index correspond to the maximal value of each one since larger values of DI and Silhouette index indicate better clustering quality.

Using Wine data set, Silhouette index with the value of 0.63 is better than Dunn since the larger value of both Dunn and Silhouette indicates better clustering quality as shown in Table 1.

In Fig. 2, we notice that the optimal number of clusters using Wine data set is corresponding to the minimal value of DB and DVI index which are 0.4 and 0.11. However, in Fig. 3 the optimal number of clusters corresponds to the maximum value of Silhouette and Dunn index which are: 0.63 and 0.56.

Fig. 2 Variation of DB and DVIndex

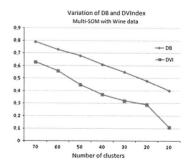

Fig. 3 Variation of Dunn
and Silhouette index

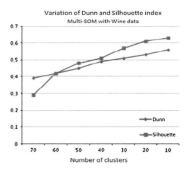

Thus, we might simply conclude that DVI is more efficient than DB index and silhouette is more efficient than Dunn index.

6 Conclusion

Classical clustering methods are developed by [22] to test 30 different validity indexes using R language.

Different clustering validity indexes are needed to assess the quality of clusters on each SOM grid. Compared with other classical clustering methods, multi-SOM is more efficient for the determination of the optimal number of clusters.

It could be applied to a wide variety of high dimensional data sets such as medical and banking data.

As a future work we will apply multi-SOM algorithm for Market Segmentation.

References

1. Kohonen, T.: Automatic formation of topological maps of patterns in a self-organising system. In: Proceedings of the 2SCIA, Scandinavian Conference on Image Analysis, pp. 214–220 (1981)
2. Lamirel, J.C.: Using artificial neural networks for mapping of science and technology: a multi self-organizing maps approach. Scientometrics **51**, 267–292 (2001)
3. Ghouila, A., Yahia, S.B., Malouche, D., Jmel, H., Laouini, D., Guerfali, Z., Abdelhak, S.: Application of multisom clustering approach to macrophage gene expression analysis. Infect. Genet. Evol. **9**, 328–329 (2008)
4. Guha, S., Rastogi, R., Shim, K.: Cure: an efficient data clustering method for very large databases. In: SIGMOD, Proceedings ACM SIGMOD international conference on Management of data, pp. 73–84. ACM Press (1998)
5. Zhang, T., Ramakrishna, R., Livny, M.: Birch: an efficient data clustering method for very large databases, pp. 103–114 (1996)
6. Karypis, G., (Sam)Han, E.-H., Kumar, V.: Chameleon: hierarchical clustering using dynamic modeling. IEEE Xplore **32**, 68–75 (1999)

7. MacQueen, J.: Some methods for classification and analysis of multivariate observations. Proc. Fifth Berkeley Symp. Math. Stat. Probab. **1**, 281–289 (1967)
8. Kaufman, L., Rousseeuw, P.: Methods clustering by means of medoids. Statistical Data Analysis Based on the L1-Norm and Related, pp. 405–417. Amsterdam, New York (1987)
9. Ester, M., Kriegel, H.-P., Sander, J., Xu, X.: A density-based algorithm for discovering clusters in large spatial databases with noise. In: Proceedings of 2nd International Conference on KDD, Portland, Oregon, pp. 226–231 (1996)
10. Ankerst, M., Breunig, M., Kriegel, H.-P., Sander, J.: Optics: ordering points to identify the clustering structure. In: SIGMOD Proceedings ACM SIGMOD International Conference on Management of Data, June 1–3, Philadelphia, Pennsylvania, USA, vol. 28, pp. 49–60. ACM Press (1999)
11. Lamirel, J.C.: Multisom: a multimap extension of the SOM model. Appl. Inf. Discov. Iconogr. Context **3**, 1790–1795 (2002)
12. Lamirel, J.C., Shehabi, S.: Multisom: a multimap extension of the SOM model. Application to information discovery in an iconographic context. In: IEEE Conference Publications, pp. 42–54 (2006)
13. Lamirel, J.C., Shehabi, S., Hoffmann, M., Francois, C.: Intelligent patent analysis through the use of a neural network: experiment of multi-viewpoint analysis with the multisom model, pp. 7–23 (2003)
14. Khanchouch, I., Boujenfa, K., Limam, M.: An Improved multi-SOM algorithm. Int. J. Netw. Secur. Appl. (IJNSA) **5**(4), 181–186 (2013)
15. Fort, J.-C., Letremy, P., Cottrell, M.: Batch Kohonen algorithm. pp. 223–230 (2002)
16. Halkidi, M., Vazirgiannis, M., Batistakis, Y.: Quality scheme assessment in the clustering process. In: Proceedings of PKDD (Principles and Practice of Knowledge Discovery in Databases), Lyon, France (2000)
17. Davies, D.L., Bouldin, D.W.: A cluster separation measure. IEEE Trans. Pattern Anal. Mach. Intell. **1**, 224–227 (1979)
18. Dunn, J.C.: A fuzzy relative of the ISODATA process and its use in detecting compact well-separated clusters. Cybern. Syst. **3**, 32–57 (1974)
19. Shen, J., Chang, S.I., Lee, E.S., Deng, Y., Brown, S.J.: Determination of cluster number in clustering microarray data. Appl. Math. Comput. **169**(2), 1172–1185 (2005)
20. Rousseeuw, P.J.: Silhouettes: a graphical aid to the interpretation and validation of cluster analysis. Comput. Appl. Math. **20**, 53–65 (1987)
21. https://archive.ics.uci.edu/ml/datasets/Wine
22. Charrad, M., Ghazzali, N., Boiteau, V., Niknafs, A.: NbClust: an R package for determining the relevant number of clusters in a data set. J. Stat. Softw. **61**(6), 1–36 (2014). http://www.jstatsoft.org/v61/i06/
23. R Core Team. R: a language and environment for statistical computing. R Foundation for Statistical Computing, Vienna, Austria (2014). http://www.R-project.org/

Conformance Testing for Timed Recursive Programs

Hana M'Hemdi, Jacques Julliand, Pierre-Alain Masson
and Riadh Robbana

Abstract This paper is about conformance testing of timed pushdown automata with inputs and outputs (TPAIO), that specify both stack and clock constraints. TPAIO are used as a model for timed recursive programs. This paper proposes a novel method of off-line test generation from deterministic TPAIO. In this context, a first problem is to resolve the clock constraints. It is solved by computing a deterministic timed pushdown tester with inputs and outputs (TPTIO), that is a TPAIO with only one clock, and provided with a location fail. To generate test cases from a TPTIO, we compute from it a finite reachability automaton (RA), that relates any of its transitions to a path of the TPTIO. The RA computation takes the TPTIO transitions as a coverage criterion. The transitions of the RA, thus the paths of the TPTIO are used for generating test cases that aim at covering the reachable locations and transitions of the TPAIO.

Keywords Timed automata · Timed pushdown automata · Conformance testing · Approximated determinization · Analog-clock testing · Reachability automaton

H. M'Hemdi (✉) · J. Julliand · P.-A. Masson
FEMTO-ST/DISC, University of Franche-Comté, 16, Route de Gray,
25030 Besançon, France
e-mail: hana.mhemdi@femto-st.fr

J. Julliand
e-mail: jacques.julliand@femto-st.fr

P.-A. Masson
e-mail: pierre-alain.masson@femto-st.fr

H. M'Hemdi
LIP2, University of Tunis El Manar, Tunis, Tunisia

R. Robbana
LIP2 and INSAT-University of Carthage, Tunis, Tunisia
e-mail: riadh.robbana@fst.rnu.tn

© Springer International Publishing Switzerland 2016
R. Lee (ed.), *Computer and Information Science 2015*,
Studies in Computational Intelligence 614, DOI 10.1007/978-3-319-23467-0_14

203

1 Introduction

Systems are commonly modelled by means of transition systems such as finite automata, timed automata, etc. System formal verification as well as model-based test generation, that are very active research areas, rely on exploiting these models. This paper is about generating tests from the model of timed pushdown automata [1] with inputs and outputs (*TPAIO*).

Timed automata (*TA*) are equipped with a finite set of real-valued clocks. They have been introduced by Alur and Dill [2], and have become a standard for modelling real-time systems. Pushdown automata (*PA*) [3] are equipped with a stack, and can be used for modelling recursive systems. The model of *TPAIO*, by including both a stack and some clocks, can model recursive systems with inputs and outputs whose execution is in a real time context.

Test generation from *TPAIO* could apply to industrial case studies such as that of [6], that defines automatic synthesis of robust and optimal controllers. These kinds of controllers operate on variables that are constantly growing in real-time, such as oil pressure etc. As shown in [16], this system can be modelled as a recursive timed automaton with a safety critical objective. Benerecetti et al. [4] argues that timed recursive state machines are related to an extension of pushdown timed automata, where an additional stack, coupled with the standard control stack, is used to store temporal valuations of clocks. Therefore, the system of [6] can be modelled by means of a *TPAIO*. Also, *TPAIO* can serve as a model for the verification of real-time distributed systems, as quoted in [4].

We propose in this paper an approach for computing offline tests from a *TPAIO* model. Offline tests, contrarily to online tests that are dynamically computed along an execution, are first extracted out of the model as a set of abstract executions, to be subsequently executed on the system. We focus in this paper on testing recursive deterministic programs without inputs.

We propose a new approach for conformance testing of *TPAIO*, aiming at covering its reachable locations and transitions. The idea is to deal successively with the clock and stack constraints that could prevent a location from being reachable. Location reachability in *TPAIO* is decidable [1, 5], and time exponential [7].

Our first contribution is the construction of a timed pushdown tester with inputs and outputs (*TPTIO*) of the *TPAIO*, by adapting the determinization method of [12] (for timed IO automata with no stack) to the *TPAIO* case. Even in the restricted framework of deterministic programs, this step is useful as it produces a model with one single clock reset after each transition, in which locations reachability has been verified w.r.t. the satisfiability of the clock constraints. Additionally, this will facilitate a future extension of the method to the case of non-deterministic *TPAIO*. Fail verdicts are added as special locations to this determinized *TPAIO*: they model the observation of timeouts or of unspecified stack and output actions.

Locations reachability has further to be verified w.r.t. the stack constraints. Finkel et al. propose in [10] a polynomial method for checking locations reachability in a *PA*. It relies on a set of rules that compute, in the shape of a finite automaton, a reachability automaton (*RA*) from a *PA*. As a second contribution we propose, following them, to adapt the rules to the *TPAIO* case, with a transition coverage criterion. We compute an *RA* whose transitions are all related to one path of the *TPTIO*. The paths are used to extract a set of tests out of the original *TPAIO*: we expand the paths that reach a final location from an initial one with an empty stack. This computes a set of test cases from a *TPAIO*.

To summarize, our contributions are to: (i) define *tpioco*: a conformance relation for the *TPAIO* model; (ii) adapt the determinization method of [12] to obtain a *TPTIO*; (iii) compute an *RA* where any transition is labelled with a path of a *TPTIO*, by adapting the reachability computation of [10]; (iv) generate test cases by covering the reachable locations and transitions of the *TPAIO*. To our knowledge these problems, solved for the TA [12] and PA [10], have not been handled for the *TPAIO* yet.

The paper is organized as follows. Section 2 presents the *TA* model and the timed input-output conformance relation *tioco*. Our model of *TPAIO* is presented in Sect. 3, together with a *TPAIO* conformance relation. Our *TPAIO* test generation method is presented in Sect. 4. We discuss the soundness, incompleteness and coverage of our method in Sect. 5. We conclude and indicate future works in Sect. 6.

2 Background

This section defines *TA* and a timed input-output conformance relation.

2.1 Timed Automata

Let $Grd(X)$ be the language of clock guards defined as conjunctions of expressions $x \sharp n$ where x is a clock of X, n is a natural integer constant and $\sharp \in \{<, \leq, >, \geq, =\}$. Let $CC(X)$ be the language of clock constraints defined as conjunctions of expressions $e \sharp n$ where e is either x, $x - x$ or $x + x$.

Definition 1 (*Timed Automaton*) A *TA* is a tuple $T = \langle L, l_0, \Sigma, X, \Delta, F \rangle$ where L is a finite set of locations, l_0 is an initial location, Σ is a finite set of labels, X is a finite set of clocks, $F \subseteq L$ is a set of accepting locations and $\Delta \subseteq L \times \Sigma \times Grd(X) \times 2^X \times L$ is a finite set of transitions.

A transition is a tuple (l, a, g, X', l') denoted by $l \xrightarrow{a,g,X'} l'$ where $l, l' \in L$ are respectively the source and target locations, a ($\in \Sigma$) is an action symbol, X' ($\subseteq X$) is a set of resetting clocks and g is a guard. The operational semantics of a *TA* T is an infinite transition system $\langle S^T, s_0^T, \Delta^T \rangle$ where the states of S^T are pairs $(l, v) \in L \times (X \to \mathbb{R}^+)$, with l a location and v a clock valuation. s_0^T is the initial state and Δ^T is the set of transitions. There are two kinds of transitions in Δ^T: timed and discrete. Timed transitions are in the shape of $(l, v) \to^\delta (l, v + \delta)$ where $\delta \in \mathbb{R}^+$

is a delay, so that $v + \delta$ is the valuation v with each clock augmented by δ. Discrete transitions are in the shape of $(l, v) \rightarrow^a (l', v')$ where $a \in \Sigma$ and $(l, a, g, X', l') \in \Delta$, and such that v satisfies g and $v' = v[X' := 0]$ is obtained by resetting to zero all the clocks in X' and leaving the others unchanged. A path π of a *TA* is a finite sequence of its transitions: $l_0 \xrightarrow{a_0, g_0, X_0} l_1 \xrightarrow{a_1, g_1, X_1} l_2 \cdots l_{n-1} \xrightarrow{a_{n-1}, g_{n-1}, X_{n-1}} l_n$. A run of a *TA* is a path of its semantics. $\sigma = (l_0, v_0) \rightarrow^{\delta_0} (l_0, v_0 + \delta_0) \rightarrow^{a_0} (l_1, v_1) \rightarrow^{\delta_1} (l_1, v_1 + \delta_1) \rightarrow^{a_1} (l_2, v_2) \rightarrow^{\delta_2} \ldots \rightarrow^{a_{n-1}} (l_n, v_n)$ where $\delta_i \in \mathbb{R}^+$ and $a_i \in \Sigma$ for each $0 \leqslant i \leqslant n - 1$ is a run of π if $v_i \models g_i$ for $0 \leqslant i < n$. A run alternates timed and discrete transitions. Its trace is a finite sequence $\rho = \delta_0 a_0 \delta_1 a_1 \ldots \delta_n a_n$ of $(\Sigma \cup \mathbb{R}^+)^*$. We denote $RT(\Sigma)$ the set of finite traces $(\Sigma \cup \mathbb{R})^*$ on Σ. $P_{\Sigma_1}(\rho)$ is the projection on $\Sigma_1 \subseteq \Sigma$ of a trace ρ with the delays preserved. For example, if $\rho = 5a4b2$, then, $P_{\{a\}}(\rho) = 5a42$ i.e. $5a6$. $Time(\rho)$ is the sum of all the delays of ρ. For example, $Time(5a42) = 11$. $s_0^T \rightarrow^\rho s$ means that the state s is reachable from the initial state s_0^T, i.e. there exists a run σ from s_0^T to s whose trace is ρ. $s_0^T \rightarrow^\rho$ means that there exists s' such that $s_0^T \rightarrow^\rho s'$.

Timed Automata with Inputs and Outputs (*TAIO*) extend the *TA* model by distinguishing between input and output actions. A *TAIO* is a tuple $\langle L, l_0, \Sigma_{in} \cup \Sigma_{out} \cup \{\tau\}, X, \Delta, F \rangle$ where Σ_{in} is a set of input actions, Σ_{out} is a set of output actions and τ is an internal and unobservable action. This model is widely used in the domain of test. It models the controllable ($\in \Sigma_{in}$) and observable ($\in \Sigma_{out}$) interactions between the environment and the system. The environment, thus the tester, sends commands of Σ_{in} and observes outputs of Σ_{out}. The implementation under test (*IUT*), sends observable actions of Σ_{out} and accepts commands of Σ_{in}.

Let $\Sigma = \Sigma_{in} \cup \Sigma_{out}$ and $\Sigma_\tau = \Sigma \cup \{\tau\}$. A *TAIO* is deterministic if for all locations l in L, for all actions a in Σ_τ and for all couples of distinct transitions $t_1 = (l, a, g_1, X_1, l_1)$ and $t_2 = (l, a, g_2, X_2, l_2)$ in Δ then $g_1 \wedge g_2$ is not satisfiable. It is observable if no transition is labelled by τ. $Reach(T) = \{s^T \in S^T \mid \exists \rho \cdot (\rho \in RT(\Sigma) \wedge s_0^T \rightarrow^\rho s^T\}$ denotes the set of reachable states of a *TAIO T*. A *TAIO T* is non blocking if $\forall (s, \delta) \cdot (s \in Reach(T) \wedge \delta \in \mathbb{R}^+ \Rightarrow \exists \rho \cdot (\rho \in RT(\Sigma_{out} \cup \{\tau\}) \wedge Time(\rho) = \delta \wedge s \rightarrow^\rho))$. A *TAIO* is called input-complete if it accepts any input at any state.

2.2 Timed Input-Output Conformance Relation *tioco*

We first present the conformance theory for timed automata based on the conformance relation *tioco* [12]. *tioco* is an extension of the *ioco* relation of Tretmans [15]. The main difference is that *ioco* uses the notion of quiescence, contrarily to *tioco* in [12] where the timeouts are explicitly specified. The assumptions are that the specification of the *IUT* is a non-blocking *TAIO*, and the implementation is a non-blocking and input-complete *TAIO*. This last requirement ensures that the execution of a test case on the *IUT* does not block the verdicts to be emitted.

To present the conformance relation for a *TAIO* $T = \langle L, l_0, \Sigma_{in} \cup \Sigma_{out} \cup \{\tau\}, X, \Delta, F \rangle$, we need to define the following notations in which $\rho \in RT(\Sigma_{in} \cup \Sigma_{out})$:

- *T after* $\rho = \{s \in S^T \mid \exists \rho' \cdot (\rho' \in RT(\Sigma_\tau) \wedge s_0^T \xrightarrow{\rho'} s \wedge P_\Sigma(\rho') = \rho)\}$ is the set of states of T that can be reached by a trace ρ' whose projection $P_\Sigma(\rho')$ on the controllable and observable actions is ρ.
- *ObsTTraces*$(T) = \{P_\Sigma(\rho) \mid \rho \in RT(\Sigma_\tau) \wedge s_0^T \xrightarrow{\rho}\}$ is the set of observable timed traces of a *TAIO* T.
- *elapse*$(s) = \{\delta \mid \delta > 0 \wedge \exists \rho \cdot (\rho \in RT(\{\tau\}) \wedge Time(\rho) = \delta \wedge s \xrightarrow{\rho})\}$ is the set of delays that can elapse from the state s with no observable action.
- *out*$(s) = \{a \in \Sigma_{out} \mid s \xrightarrow{a}\} \cup elapse(s)$ is the set of outputs and delays that can be observed from the state s.

Definition 2 (*tioco*) Let $T = (L, l_0, \Sigma_\tau, X, \Delta, F)$ be a specification and $I = (L^I, l_0^I, \Sigma_\tau^I, X^I, \Delta^I, F^I)$ be an implementation of T. Formally, I conforms to T, denoted I *tioco* T iff $\forall \rho \cdot (\rho \in ObsTTraces(T) \implies out(I \text{ after } \rho) \subseteq out(T \text{ after } \rho))$.

It means that the implementation I conforms to the specification T if and only if after any timed trace enabled in T, each output or delay of I is specified in T.

3 Model and Conformance Relation

This section defines our model: *TPAIO*, as well as *tpioco*, a conformance relation for *TPAIO*. We show an example of a *TPAIO* that models a recursive program.

3.1 Timed Pushdown Automata with Inputs and Outputs

A *TPAIO* $T = \langle L, l_0, \Sigma, \Gamma, X, \Delta, F \rangle$ is a *TAIO* equipped with a stack. Its operational semantics is a transition system $< S^T, s_0^T, \Delta^T >$ where the locations –called states– are configurations made of three components (l, v, p) with l a location of the *TPAIO*, v a clock valuation in $X \to \mathbb{R}^+$ and p a stack content in Γ^*.

Definition 3 (*TPAIO*) A *TPAIO* is a tuple $\langle L, l_0, \Sigma, \Gamma, X, \Delta, F \rangle$ where L is a finite set of locations, l_0 is an initial location, $\Sigma = \Sigma_{in} \cup \Sigma_{out} \cup \{\tau\}$ where Σ_{in} is a finite set of input actions, Σ_{out} is a finite set of output actions and $\{\tau\}$ is an internal and unobservable action, Γ is a stack alphabet ($\Sigma_{out} \cap \Sigma_{in} = \emptyset$, $\Sigma_{in} \cap \Gamma = \emptyset$ and $\Sigma_{out} \cap \Gamma = \emptyset$), X is a finite set of clocks, $F \subseteq L$ is a set of accepting locations, $\Delta \subseteq L \times (\Sigma_{in} \cup \Sigma_{out} \cup \Gamma^{+-}) \times Grd(X) \times 2^X \times L$ is a finite set of transitions where $\Gamma^{+-} = \{a^+ \mid a \in \Gamma\} \cup \{a^- \mid a \in \Gamma\}$.

The symbols of Γ^{+-} represent either a push operation (of the symbol a) denoted a^+, or a pop operation denoted a^-. A transition is a tuple (l, a, g, X', l') denoted by $l \xrightarrow{a,g,X'} l'$ where $l, l' \in L$ are respectively the source and target locations, $a \in \Sigma \cup \Gamma^{+-}$ is either a label or a stack action, X' ($\subseteq X$) is a set of resetting clocks and g is a guard. There are two kinds of transitions in the semantics, timed and discrete. Timed transitions are in the shape of $(l, v, p) \rightarrow^\delta (l, v + \delta, p)$. For a transition (l, act, g, X', l'), there are three types of discrete transitions when v satisfies g: (1) *push* when $act = a^+$: $(l, v, p) \rightarrow^{a^+} (l', v[X' := 0], p.a)$ where $a \in \Gamma$, (2) *pop* when $act = a^-$: $(l, v, p.a) \rightarrow^{a^-} (l', v[X' := 0], p)$ where $a \in \Gamma$, (3) *output, input or internal* when $act = A \in \Sigma$: $(l, v, p) \rightarrow^A (l', v[X' := 0], p)$. A *TPAIO* is normalized if it executes separately push and pop operations. Any *TPAIO* can be normalized since any *PA* can be normalized [14]. Due to the class of application, we consider in the remainder of the paper that the *TPAIO* are normalized deterministic timed pushdown automata with outputs and without inputs. We denote a the actions of Γ and A the actions of Σ_{out}.

We define *tpioco*, our *TPAIO* conformance relation, as an extension of the *tioco* conformance relation [12]. It is the same relation as *tioco* for *TAIO* by considering the whole alphabet to be $\Sigma_{out} \cup \Gamma^{+-} \cup \{\tau\}$ instead of $\Sigma_{in} \cup \Sigma_{out} \cup \{\tau\}$. The output alphabet is $\Sigma_{out} \cup \Gamma^{+-}$ instead of Σ_{out} and there is no input alphabet.

3.2 Modelling of Recursive Programs

Figure 1 shows a program that recursively computes the nth Fibonacci number, with instructions labels from l_0 to l_6. We abstract the control flow graph of a recursive program by a *PA*, as explained in [9]. Figure 2 shows a *TPAIO* that abstracts the program of Fig. 1. Here the time constraints have been added arbitrarily for illustration purposes. The location labels are the instruction labels in the program of Fig. 1. Fib_1^+ and Fib_2^+ are respectively the (push) calls $Fib(n - 1)$ and $Fib(n - 2)$. Fib_1^- and Fib_2^- are respectively the (pop) returns from the calls $Fib(n - 1)$ and $Fib(n - 2)$. Thus $\Gamma = \{Fib_1, Fib_2\}$. Such an example is a transformational system in which the tester observes any action of the program. Therefore, we choose that all the executions of atomic instructions and conditions are in Σ_{out}. They are labelled from A to E as

Fig. 1 A fibonacci computation program $Fib(n)$

```
int  Fib(int n)
    l0 : int res1, res2;
    if l1 : n ⩽ 1 then
        l2 : return n;
    else
        l3 : res1 = Fib(n − 1); //Fib1+
        l4 : res2 = Fib(n − 2); //Fib2+
        l5 : return res1 + res2;
    fi
    l6 : end.
```

Fig. 2 A *TPAIO* modelling
the program of Fig. 1

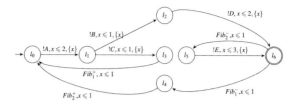

follows: $A \stackrel{def}{=} \text{int } res_1, res_2, B \stackrel{def}{=} n \leqslant 1, C \stackrel{def}{=} n > 1, D \stackrel{def}{=} return \ n$ and $E \stackrel{def}{=} return$
$res_1 + res_2$. We use the notation *!act* to denote the output action *act*.

4 Test Generation from *TPAIO*

This section presents our test generation method from a deterministic *TPAIO*. We
first present the test generation process and then the three steps of our method.

4.1 Test Generation Process

The data flow diagram of Fig. 3 shows the three steps of the test generation process
that we propose in this paper:

1. Construction of a *TPTIO* of a *TPAIO*: A *TPAIO* specifies clock constraints. For
   this reason, we propose to compute a Timed Pushdown Tester with Inputs and
   Outputs (*TPTIO*) that resolves the clock constraints. The tester obtained is a
   TPAIO with one clock, reset each time the tester observes an action, and provided
   with a location *fail*.
2. Computation of an *RA* from the *TPTIO*: pop actions depend on the content of the
   stack. This step computes one or many paths between two symbolic locations of
   a *TPTIO* by respecting the stack constraints, i.e. such that the stack is empty in

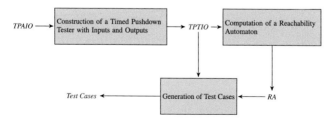

Fig. 3 Test generation process from a *TPAIO*

the target location. The *RA* is a finite automaton with any of its transition related to one path of the *TPTIO*. Such a transition is called a π-transition.

3. Generation of test cases as correct behaviours of the *TPAIO*, computed from the *TPTIO*. There are two sub-steps: (a) generation of test paths of π-transitions that go from an initial to a final location of the *RA*; (b) generation of *TPTIO* test cases, by adding to the test paths the location *fail* and the transitions that lead to it.

4.2 Construction of a TPTIO from a TPAIO

In [12], Krichen and Tripakis propose a method for conformance testing of non-deterministic *TAIO*. They propose an algorithm for generating test cases. They compute a tester that has only one clock, reset each time the tester observes an action.

We propose to adapt the method of [12] for computing a *TPTIO* from a *TPAIO*. Let *out*(*l*) be the set of transitions leaving the location *l* in the *TPAIO*. A verdict *fail* is emitted if either an unspecified stack or output action is observed, or a stack or output actions of *out*(*l*) is observed earlier or later than specified, or a timeout occurs. A *TPTIO* has only one clock which is reset every time the tester observes an action. As a consequence, all the guards of a *TPTIO* are satisfiable. We define a *TPTIO* of a *TPAIO* in Definition 4.

Definition 4 (*TPTIO*) The *TPTIO* $T^T = (L^T, l_0^T, \Sigma_{out}, \Gamma, \{y\}, \Delta^T, F^T)$ of a *TPAIO* $T = \langle L, l_0, \Sigma_{out} \cup \{\tau\}, \Gamma, X, \Delta, F \rangle$ is a *TPAIO* with only one clock y that is a new clock w.r.t X where:

- $L^T \subseteq (L \times CC(X \cup \{y\})) \cup \{fail\})$ is a set of symbolic locations,
- l_0^T is the initial symbolic location,
- $F^T \subseteq L^T$ is a set of accepting symbolic locations,
- $\Delta^T \subseteq L^T \times \Sigma_{out} \cup \Gamma^{+-} \times Grd(\{y\}) \times \{y\} \times L^T$ is a finite set of transitions.

The computation, taken from [12], of a partition where each part is in $Grd(\{y\})$ is as follows: let K be the greatest constant appearing in a constraint of a given symbolic location l^T or in a guard of a given transition of T. The following set of intervals is a partition: $\{[0, 0],]0, 1[, [1, 1],]1, 2[, \ldots, [K, K],]K, \infty[\}$. Before presenting the method to compute Δ^T, we need to define the following sets:

- $\Delta_a = \{t \mid t \in \Delta \wedge \exists (l, g, X', l').(t = (l, a, g, X', l'))\}$ is the set of transitions of Δ labelled by a.
- $\Delta_a((l, v), u) = \{(l, a, g, X', l') \in \Delta_a \mid v \wedge u \wedge g \text{ satisfiable}\}$ is the set of transitions labelled by a whose guards are satisfied by (l, v) where v is in $CC(X \cup \{y\})$ and the clock y is equal to u.

Since our model is that of deterministic *TPAIO*, Δ_a and $\Delta_a(l^T, u)$ contain at most one transition. For all intervals u, the coarsest partition is obtained from l^T by taking the union of the intervals that have the same set $\Delta_a(l^T, u)$. For a symbolic location $l^T \in L^T$ of T^T, $a \in \Sigma_{out} \cup \Gamma^{+-}$:

- $usucc(l^T) = l^{T'}$ such that $\exists \rho.(\rho \in RT(\{\tau\}) \wedge l^T \rightarrow^\rho l^{T'})$ is the symbolic location reachable from l^T by applying a sequence of unobservable actions.
- $dsucc(l^T, a) = l^{T'}$ such that $l^T \rightarrow^a l^{T'}$ is the symbolic location reachable from l^T by applying the action a.

In the initial location l_0^T, all the clocks equal zero, including y. The construction of the *TPTIO* repeats the following step: selection of a symbolic location $l^T \in L^T$ and application of the following possibilities to add new transitions to Δ^T: **(i)** output and stack actions: for every action $a \in \Sigma_{out} \cup \Gamma^{+-}$, for every coarsest partition u, if $\Delta_a(l^T, u) = \emptyset$ then the transition $(l^T, a, u, \{y\}, fail)$ is added to Δ^T. Otherwise the transition $(l^T, a, u, \{y\}, usucc(dsucc(l^T \cap u, a)))$ is added to Δ^T; **(ii)** timeout: let K be the greatest constant appearing in the constraint of l^T or in a guard of the transitions leaving l^T. Then, the transition $(l^T, -, y > K, \{y\}, fail)$ is added to Δ^T.

The first symbolic location selected is l_0^T. Adding new transitions to a *TPTIO* implies adding new symbolic locations to the *TPTIO*. The algorithm terminates when all the new symbolic locations are selected and treated.

Notice that the number of locations of the *TPTIO* could scale exponentially with that of the *TPAIO*. However, the impact of this blowup can be limited by putting time constraints on blocks of instructions rather than on single instructions.

Example 1 Figure 4 shows the *TPTIO* of the *TPAIO* of Fig. 2 where $v_i \overset{def}{=} 0 \leqslant x - y \leqslant i$. The label $a_1|a_2|\ldots|a_n$ denotes the set of labels $\{a_1, a_2,\ldots, a_n\}$. $F = Fib_1^+|Fib_1^-|Fib_2^+|Fib_2^-|?A|?B|?C|?D|?E, F0 = F\backslash\{?A\}, F1 = F\backslash\{?B, ?C\}, F2 = F\backslash\{?D\}, F3 = F\backslash\{Fib_1^+\}, F4 = F\backslash\{Fib_2^+\}, F5 = F\backslash\{?E\}$ and $F6 = F\backslash\{Fib_1^-, Fib_2^-\}$.

4.3 Reachability Automaton of a TPTIO Computation

A *TPAIO* does not only specify clock constraints but also stack constraints. Therefore, applying the algorithm of [12] for generating analog-clock tests from *TPAIO* is not sufficient. It is necessary, for avoiding system deadlocks, to additionally take the stack content into account. For example, the pop action of a symbol that would not be on top of the stack would provoke a deadlock. We compute a reachability automaton for taking the stack constraints into account.

Let $(L^T, l_0^T, \Sigma_{out}, \Gamma, \{y\}, \Delta^T, F^T)$ be a *TPTIO*. We propose to compute a representation of its reachable locations from its initial location. This representation is called the *reachability automaton* of the *TPTIO*. It is a finite automaton whose transition labels are sequences of transitions of the *TPTIO* ($\in \Delta^{T^*}$). A π-transition $(l^T, \pi, l^{T'})$ is a transition that reaches the symbolic location $l^{T'}$ from l^T and leaves the stack unchanged at the end, by taking the path π. We propose in Definition 5 the rules R_1 to R_4 that, applied repeatedly, define the *RA*.

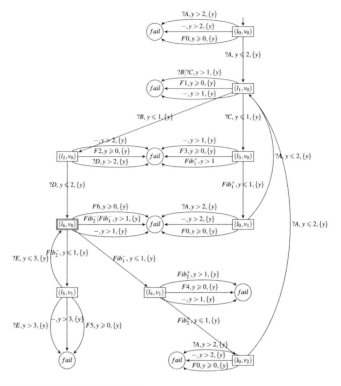

Fig. 4 The *TPTIO* of the *TPAIO* of Fig. 2

Definition 5 (*RA of a TPTIO*) The *RA* of a *TPTIO* $(L^T, l_0^T, \Sigma_{out}, \Gamma, \{y\}, \Delta^T, F^T)$ is the automaton $(L^R, l_0^T, (\Delta^T)^*, \Delta^R, F^T)$ where $L^R = L^T \setminus \{fail\}$ and $\Delta^R \subseteq L^R \times (\Delta^T)^* \times L^R$ is the smallest relation that satisfies the following conditions. Let $t_1 \stackrel{def}{=} (l_1^T, a^+, g_1, \{y\}, l_2^T), t_2 \stackrel{def}{=} (l_2^T, a^-, g_2, \{y\}, l_3^T)$ and $t_3 \stackrel{def}{=} (l_3^T, a^-, g_3, \{y\}, l_4^T)$ be three transitions in Δ^T where l_2^T, l_3^T, l_4^T differ from the location *fail*.

- **R₁**: $(l^T, t, l^{T'}) \in \Delta^R$ if $t \stackrel{def}{=} (l^T, A, g, \{y\}, l^{T'})$ and $t \in \Delta^T$,
- **R₂**: $(l_1^T, t_1.t_2, l_3^T) \in \Delta^R$,
- **R₃**: $(l_1^T, t_1.\pi.t_3, l_4^T) \in \Delta^R$ if $(l_2^T, \pi, l_3^T) \in \Delta^R$ and t_1 is not a prefix of π or t_3 is not a suffix of π.
- **R₄**: $(l_1^T, \pi_1.\pi_2, l_3^T) \in \Delta^R$ if $(l_1^T, \pi_1, l_2^T) \in \Delta^R$ and $(l_2^T, \pi_2, l_3^T) \in \Delta^R$ and π_1 is not a prefix of π_2 and π_2 is not a suffix of π_1.

We have adapted an algorithm by Finkel et al. [10], that originally computes an *RA* from a *PA*, to compute an *RA* from a *TPTIO*. Our modifications are as follows:

1. we compute the path of each transition. Any π-transition in the *RA* corresponds to the path π in the *TPTIO*;

2. the problem addressed in [10] being to check the location reachability, the paths are ignored and there is only one transition between two symbolic locations l^T and $l^{T'}$. We record as many transitions as required to cover all the transitions between the two symbolic locations;

3. we consider, by means of the rule R_1, the transitions that emit a symbol of Σ_{out} in addition to the push and pop ones;

4. the reflexive transitions are not used in [10] because they don't cover any new state. By contrast in the rule R_4, we possibly extend an existing π-transition on its right or on its left by one occurrence of a reflexive transition, provided that it covers at least one new transition.

The computation of the paths is based on transition coverage. Adding a new π-transition $(l^T, \pi, l^{T'})$ is performed only if π covers a new transition between l^T and $l^{T'}$ of the *TPTIO*. Thus our algorithm applies the rules R_1 to R_4 to compute the smallest set of transitions Δ^R that covers all the transitions.

Example 2 There are two transitions $((l_0, v_0), \pi, (l_6, v_0))$ that go from the initial symbolic location (l_0, v_0) to the accepting symbolic location (l_6, v_0):

1. $\pi \stackrel{def}{=} (l_0, v_0) \xrightarrow{A, y \leqslant 2, \{y\}} (l_1, v_0) \xrightarrow{B, y \leqslant 1, \{y\}} (l_2, v_0) \xrightarrow{D, y \leqslant 2, \{y\}} (l_6, v_0)$

2. $\pi \stackrel{def}{=} (l_0, v_0) \xrightarrow{A, y \leqslant 2, \{y\}} (l_1, v_0) \xrightarrow{C, y \leqslant 1, \{y\}} (l_3, v_0) \xrightarrow{Fib_1^+, y \leqslant 1, \{y\}} (l_0, v_1) \xrightarrow{A, y \leqslant 2, \{y\}}$
   $(l_1, v_0) \xrightarrow{B, y \leqslant 1, \{y\}} (l_2, v_0) \xrightarrow{D, y \leqslant 2, \{y\}} (l_6, v_0) \xrightarrow{Fib_1^-, y \leqslant 1, \{y\}} (l_4, v_1) \xrightarrow{Fib_2^+, y \leqslant 1, \{y\}}$
   $(l_0, v_2) \xrightarrow{A, y \leqslant 2, \{y\}} (l_1, v_0) \xrightarrow{B, y \leqslant 1, \{y\}} (l_2, v_0) \xrightarrow{D, y \leqslant 2, \{y\}} (l_6, v_0) \xrightarrow{Fib_2^-, y \leqslant 1, \{y\}}$
   $(l_5, v_1) \xrightarrow{E, y \leqslant 3, \{y\}} (l_6, v_0)$.

4.4 Generation of Correct Behaviour Test Cases

Definition 6 (*Test Case*) Let $T = \langle L, l_0, \Sigma_{out} \cup \{\tau\}, \Gamma, X, \Delta, F \rangle$ be a *TPAIO* specification and $T^T = (L^T, l_0^T, \Sigma_{out}, \Gamma, \{y\}, \Delta^T, F^T)$ be the *TPTIO* of T. A test case is a deterministic acyclic *TPAIO* whose locations are either configurations (l, v, p), or *pass*, or *fail*, or *stack_fail*.

We first define a test case in Definition 6. We propose to select the executions that reach a final symbolic location with an empty stack, for producing a set of nominal test cases. For this, we select the π-transitions in *RA* that go from an initial symbolic location to a final one. Any path of each selected transition is then turned into a test case by adding, from each state it reaches, the corresponding transitions that lead to *fail* in the tester. The last state reached is a final state with empty stack. It is replaced by the verdict *pass*. The non-verdict nodes are configurations (location, clock valuation, stack content) of the semantics of the *TPAIO*. To model the case where the pop of a symbol by the *IUT* is observed, although the symbol should not be on top of the stack according to the specification, we propose to add the verdict *stack_fail*. For each state (l, v, p) in each test case, for every action $a^- \in \Gamma^-$, for

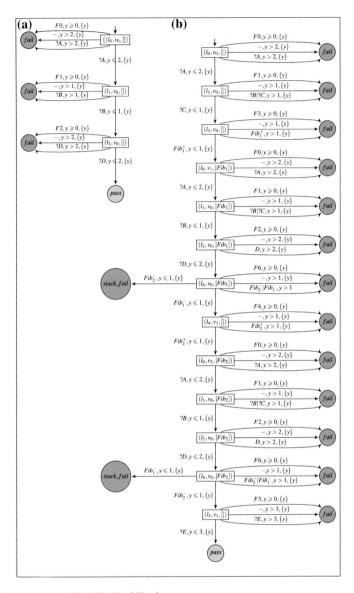

Fig. 5 Two test cases of the *TPAIO* of Fig. 2

every coarsest partition u where $\Delta_{a^-}((l, v), u) \neq \emptyset$, if the symbol a is not on the top of p, then the transition $((l, v, p), a^-, u, \{y\}, stack\_fail)$ is added to the test case. Thus, the result is a set of test cases, in which the actions are observable (the stack and output actions). Figure 5 shows the two test cases issued from the π-transition $((l_0, v_0), \pi, (l_6, v_0))$ of Example 2. For example, if the tester observes the pop of the symbol Fib_2^- from the state $(l_6, v_0, [Fib_1])$, then the verdict *stack_fail* is emitted.

The tests generated then have to be executed on the IUT. As *TPAIO* are abstractions, the tests are not in general guaranteed to be instantiable as concrete executions of the *IUT*. This is the case in our example of a recursive program whose evaluation conditions have been abstracted in the *TPAIO*. To select the concretizable test cases and to compute their inputs, we propose to use symbolic execution [11], as a mean for analysing a path and finding the corresponding program inputs. A constraint solver may also be invoked while executing a given test case [13]. The satisfiability of a constraint can be efficiently evaluated by means of SMT solvers such as Z3 [8]. If the constraint is satisfiable, then the test case is concretizable. The solver also finds a solution for the constraint of this concretizable test case. It represents the concrete inputs that lead to the execution of the test case being considered. For example, the test case of Fig. 5b represents the trace $ACFib_1^+ABDFib_1^- Fib_2^+ABDFib_2^- E$, which corresponds to the following successive instructions: int $res_1, res_2; n > 1; res_1 = Fib(n-1)$; int $res_1, res_2; n - 1 \leqslant 1$; return $n - 1$; $res_1 = Fib(n - 2)$; int $res_1, res_2; n - 2 \leqslant 1$; return $n - 2$ and return $res_1 + res_2$. It corresponds to the following path constraint: $n > 1 \wedge n - 1 \leqslant 1 \wedge n - 2 \leqslant 1$. This constraint is satisfiable and a solution is $n = 2$. Thus, this test case corresponds to $Fib(2)$. In our case, we obtain two test cases which are concretizable. The other test case (see Fig. 5a), corresponds to $Fib(0)$ or $Fib(1)$.

5 Soundness, Incompleteness and Coverage of the Method

This section discusses the soundness, incompleteness and coverage of our method for generating tests from a *TPAIO*.

5.1 Soundness

Theorem 1 *A symbolic location l_i^T is reachable with an empty stack in a TPTIO if there exists a π-transition (l_0^T, π, l_i^T) in its RA.*

Proof The proof is by induction and by cases on each rule. The induction assumption is that the *RA* transitions that are merged into new transitions are sound. We prove this assumption to be true by proving that the rules R_1 and R_2, that create *RA* transitions only from *TPTIO* ones, are sound. Then we prove that the rules R_3 and R_4 preserve that soundness.

- R_1 case: if there is a transition $(l_1, v_1) \xrightarrow{A, g_1, \{y\}} (l_2, v_2)$ $(l_2, v_2) \in \Delta^R$ then (l_2, v_2) is reachable from (l_1, v_1) in the *TPTIO* because by definition of the *TPTIO*, $v_1 \wedge g_1$ is satisfiable.

- R_2 case: if there is a transition $(l_1, v_1) \xrightarrow{a^+, g_1, \{y\}} (l_2, v_2) \xrightarrow{a^-, g_2, \{y\}} (l_3, v_3)$ (l_3, v_3) $\in \Delta^R$ then (l_3, v_3) is reachable from (l_1, v_1) in the *TPTIO* because it is always possible to pop a after a has been pushed and by definition of the *TPTIO*, $v_1 \wedge g_1$ and $v_2 \wedge g_2$ are satisfiable.

- R_3 case: if there is a transition $(l_1, v_1) \xrightarrow{a^+, g_1, \{y\}} (l_2, v_2) \xrightarrow{\pi} (l_3, v_3) \xrightarrow{a^-, g_3, \{y\}} (l_4, v_4)$
  $(l_4, v_4) \in \Delta^R$ then (l_4, v_4) is reachable from (l_1, v_1) in the *TPTIO* because (l_3, v_3) is reachable from (l_2, v_2) according to the induction hypothesis, and it is always possible to pop a after a has been pushed, and by definition of the *TPTIO* $v_1 \wedge g_1$ and $v_3 \wedge g_3$ are satisfiable.

- R_4 case: if there is a transition $(l_1, v_1) \xrightarrow{\pi_1} (l_2, v_2) \xrightarrow{\pi_2} (l_3, v_3)$ $(l_3, v_3) \in \Delta^R$ then (l_3, v_3) is reachable from (l_1, v_1) in the *TPTIO* because (l_2, v_2) is reachable from (l_1, v_1) and (l_3, v_3) is reachable from (l_2, v_2), according to the induction hypothesis.

Theorem 2 *A location l_i is reachable with an empty stack in a TPAIO iff there exists a π-transition $((l_0, v_0), \pi, (l_i, v_i))$ in the RA of its TPTIO.*

Proof Let g be the guard of a transition (l, a, g, X', l'). A transition $((l, v), a, u, \{y\},$ $usucc(dsucc((l, v) \cap u, a)))$ is added to Δ^T where $a \in \Sigma \cup \Gamma^{+-}$ only if $v \wedge u \wedge g$ is satisfiable, by definition of $\Delta_a((l, v), u)$ in Sect. 4.2. Thus the construction of a *TPTIO* from a *TPAIO* takes the clock constraints into account. It preserves both the clock and stack constraints of the *TPAIO*. Thus, Theorem 2 is a direct consequence of Theorem 1.

Our tests are correct in the sense that only non-conform executions are rejected.

Theorem 3 *Let $\pi = (l_0, v_0, p_0) \xrightarrow{a_0, g_0, \{y\}} (l_1, v_1, p_1) \xrightarrow{a_1, g_1, \{y\}} \ldots (l_{n-1}, v_{n-1},$ $p_{n-1}) \xrightarrow{a_{n-1}, g_{n-1}, \{y\}} (l_n, v_n, p_n) \xrightarrow{a_n, g_n, \{y\}} l_{n+1}$ be a path of a test case of a specification $T = \langle L, l_0, \Sigma_{out} \cup \{\tau\}, \Gamma, X, \Delta, F \rangle$ where $l_i \in L$, v_i is a clock constraint in $CC(X \cup \{y\})$, $g_i \in Grd(\{y\})$, $p_i \in \Gamma^*$, $a_i \in \Sigma_{out} \cup \Gamma^{+-} \cup \{-\}$ for each $0 \leqslant i \leqslant n$ and $l_{n+1} \in \{fail, stack\_fail\}$. If a verdict fail or stack_fail is observed while executing the implementation I, then I does not conform to the specification T.*

Proof Let $\rho = \delta_0 a_0 \delta_1 a_1 \delta_2 \ldots \delta_{n-1} a_{n-1} \delta_n a_n \in RT(\Sigma_{out} \cup \Gamma^{+-})$ be the trace of the path π. (l_n, v_n, p_n) is the current symbolic location after the execution of $\delta_0 a_0 \delta_1 a_1 \delta_2 \ldots \delta_{n-1} a_{n-1} \delta_n$ and g_n is the coarsest partition computed such that $\delta_n \in g_n$. Reaching *fail* or *stack_fail* is due to one of the following three cases:

- *fail* occurs after an unspecified stack or output action a_n has been observed, according to item (i) in Sect. 4.2. If $\Delta_{a_n}((l_n, v_n), u_n) = \emptyset$, then the transition $((l_n, v_n), a_n, \{y\}, fail)$ is a transition of the *TPTIO*. Therefore, $a_n \notin out(T \ after$ $\delta_0 a_0 \delta_1 a_1 \delta_2 \ldots \delta_{n-1} a_{n-1} \delta_n)$, and I does not conform to T.
- *fail* occurs after a timeout δ_n ($a_n = -$) has been observed, according to item (ii) in Sect. 4.2. Therefore, $v_n + \delta_n \notin out(T \ after \ \delta_0 a_0 \delta_1 a_1 \delta_2 \ldots \delta_{n-1} a_{n-1})$, and I does not conform to T.

- *stack_fail* occurs after a pop action a_n has been observed, acceptable by the specification $(\Delta_{a_n}((l_n, v_n), u_n) \neq \emptyset)$, but in a context where the symbol a should not be on top of the stack p_n. Therefore, I does not conform to T.

Thus for every non-conformance detected by a test case, there is a non-conformance between the implementation and the specification (*TPAIO*).

5.2 Incompleteness

The equivalence relation used to compute a *TPTIO* from a *TPAIO* can lead to a loss of precision. It should be possible to build more precise test cases than the ones computed by our method. Consider for example the *TPAIO* of Fig. 6a. The *TPTIO* of Fig. 6c is more precise than the one of Fig. 6b. The trace $0a^+2a^-$ leads to the symbolic location $(l_2, 0 \leqslant x - y < 4)$ in the *TPTIO* of Fig. 6b. It leads to the symbolic location $(l_2, 0 \leqslant x - y \leqslant 3)$ in the *TPTIO* of Fig. 6c, but not to the symbolic location $(l_2, 0 < x - y < 4)$. We remark that the symbolic location $(l_2, 0 \leqslant x - y \leqslant 3)$ is more precise than $(l_2, 0 \leqslant x - y < 4)$.

5.3 Coverage

In Sect. 4.3, we have presented a method for computing an *RA* from a *TPTIO*. The algorithm that computes the *RA* takes into account the coverage of the transitions of the *TPTIO*. It adds a new π-transition $(l^T, \pi, l^{T'})$ only if π covers a new transition w.r.t the other π'-transitions $(l^T, \pi', l^{T'})$. The paths of all the π-transitions that go to a final symbolic location of the *RA* cover all the transitions of the *TPTIO*. But since some test cases might be unconcretizable, the set of concrete test cases is not guaranteed to cover all the transitions of the *TPAIO*. When all the tests are concretizable (it is the case in our example), then all the reachable locations and all the reachable transitions of the *TPAIO* are covered.

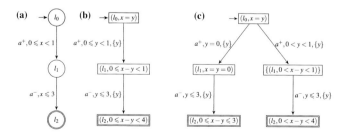

Fig. 6 Two *TPTIO* (**b**) and (**c**) of a *TPAIO* (**a**) (without *fail* location). (**c**) Is more precise than (**b**)

6 Conclusion and Further Works

We have presented a method to generate test from *TPAIO*. To our knowledge, this had not been treated before in the literature. Our method proceeds by computing reachable locations and transitions, and generating off-line tests from a deterministic *TPAIO* that models a timed recursive program. The tester observes the stack and output actions, as well as the delays. Our method first adapts the tester computation method of [12] for *TA* to the *TPAIO* case. We obtain a *TPTIO* that is a *TPAIO* with only one clock. Its locations are defined as being either rejecting (*fail*) or symbolic locations. In a second step, we adapt another algorithm presented in [10] for *PA*, for computing the *RA* of the *TPTIO*. We compute the paths of the *TPTIO* associated to each transition of the *RA*. The computation of the *RA* takes into account the coverage of all the transitions of the *TPAIO*. By using the paths of transitions of *RA* and *TPTIO*, test cases are generated. If they are concretizable, they cover all the reachable locations and transitions of the *TPAIO*.

Our work is currently for deterministic timed pushdown automata with outputs. We intend as a future work to generalize to non-deterministic timed pushdown automata with inputs and outputs, to target general timed recursive systems. Also, at this stage of our work, we have developed a proof-of-concept prototype to experimentally validate our approach, in which the test generation process have been automated. We intend as a future work to develop this tool, in order to be able to perform larger scale experiments.

References

1. Abdulla, P.A., Atig, M.F., Stenman, J.: Dense-timed pushdown automata. In: LICS, pp. 35–44 (2012)
2. Alur, R., Dill, D.L.: A theory of timed automata. TCS **126**(2), 183–235 (1994)
3. Autebert, J.M., Berstel, J., Boasson, L.: Context-free languages and pushdown automata. In: Handbook of Formal Languages, vol. 1, pp. 111–174. Springer (1997)
4. Benerecetti, M., Minopoli, S., Peron, A.: Analysis of timed recursive state machines. In: TIME 2010, pp. 61–68. IEEE (2010)
5. Bouajjani, A., Echahed, R., Robbana, R.: On the automatic verification of systems with continuous variables and unbounded discrete data structures. Hybrid Systems II, LNCS, vol. 999, pp. 64–85. Springer, Berlin (1995)
6. Cassez, F., Jessen, J.J., Larsen, K.G., Raskin, J.F., Reynier, P.A.: Automatic synthesis of robust and optimal controllers—an industrial case study. In: HSCC'09, pp. 90–104. Springer (2009)
7. Chadha, R., Legay, A., Prabhakar, P., Viswanathan, M.: Complexity bounds for the verification of real-time software. In: VMCAI'10. LNCS, vol. 5944, pp. 95–111. Springer (2010)
8. de Moura, L.M., Bjørner, N.: Z3: an efficient SMT solver. In: TACAS. LNCS, vol. 4963, pp. 337–340 (2008)
9. Dreyfus, A., Héam, P.C., Kouchnarenko, O., Masson, C.: A random testing approach using pushdown automata. STVR **24**(8), 656–683 (2014)
10. Finkel, A., Willems, B., Wolper, P.: A direct symbolic approach to model checking pushdown systems (ext. abs.). In: Infinity. ENTCS, vol. 9, pp. 27–37 (1997)
11. Godefroid, P.: Test generation using symbolic execution. In: IARCS, pp. 24–33 (2012)

12. Krichen, M., Tripakis, S.: Conformance testing for real-time systems. FMSD **34**(3), 238–304 (2009)
13. Păsăreanu, C.S., Rungta, N., Visser, W.: Symbolic execution with mixed concrete-symbolic solving. In: ISSTA'11, pp. 34–44. ACM (2011)
14. Sénizergues, G.: L(a) = l(b)? decidability results from complete formal systems. In: ICALP. LNCS, vol. 2380, pp. 1–37. Springer (2002)
15. Tretmans, J.: Testing concurrent systems: a formal approach. In: Concurrency Theory. LNCS, vol. 1664, pp. 46–65. Springer (1999)
16. Trivedi, A., Wojtczak, D.: Recursive timed automata. In: ATVA'10, LNCS, vol. 6252, pp. 306–324. Springer (2010)

Instruction Level Loop De-optimization

Loop Rerolling and Software De-pipelining

Erh-Wen Hu, Bogong Su and Jian Wang

Abstract Instruction level loop optimization has been widely used in modern compilers. Decompilation—the reverse of compilation—has also generated much interest for its applications in porting legacy software written in assembly language to new architectures, re-optimizing assembly code, and more recently, in detecting and analyzing malware. However, little work has been reported on loop decompilation at instruction level. In this paper, we report our work on loop de-optimization at instruction level. We demonstrate our approach with a practical working example and carried out experiments on TIC6x, a digital signal processor with a compiler supporting instruction level parallelism. The algorithms developed in this paper should help interested readers gain insight especially in the difficult tasks of loop rerolling and software de-pipelining, the necessary steps to decompile loops at instruction level.

Keywords Decompilation · Instruction level loop de-optimization · Loop rerolling · Software de-pipelining

Abbreviations

DSP	Digital signal processor
DDG	Data dependency graph
VLIW	Very long instruction word
ILP	Instruction level parallelism
TI	Texas Instruments

E.-W. Hu · B. Su (✉)
Department of Computer Science, William Paterson University, Wayne, NJ, USA
e-mail: sub@wpunj.edu

E.-W. Hu
e-mail: hue@wpunj.edu

J. Wang
Mobile Broadband Software Design, Ericsson, Ottawa, ON, Canada
e-mail: jian.z.wang@ericsson.com

© Springer International Publishing Switzerland 2016
R. Lee (ed.), *Computer and Information Science 2015*,
Studies in Computational Intelligence 614, DOI 10.1007/978-3-319-23467-0_15

221

1 Introduction

Decompilation techniques [8, 9] have been applied to many areas such as porting legacy software written in assembly language to new architectures, re-optimizing assembly code [1], detecting bugs [6] and malware [7]. Decompilation is a complex process typically involves operations such as unpredication and unspeculation [16], reconstructing control structures [21], resolution of branch delays [3], loop rerolling [17] and software de-pipelining [4, 5, 18].

Software pipelining [13] is a loop parallelization technique used to speed up loop execution. It is widely implemented in optimizing compilers for very long instruction word (VLIW) architecture such as IA-64, Texas Instruments (TI) C6X digital signal processors (DSP) that support instruction level parallelism (ILP). To further enhance the performance of DSP applications, software pipelining is often combined with loop unrolling [14]. Therefore, it is often necessary to perform both loop rerolling and software de-pipelining in order to de-optimize loops at instruction level.

Recently Kroustek investigated the decompilation of VLIW executable files and presented the decompression of VLIW assembly code bundles [11]. However the paper did not address the de-optimization of the code at instruction level. In general, loop de-optimization is much more difficult at instruction level than at higher levels because processors that support ILP tend to have more complicated architectures and instruction sets. Furthermore, compilers for these processors often apply various optimization techniques during different phases of compilation in order to better utilize the ILP features of the processors. For example, TIC6x DSP processor contains two data paths and each of which consists of four functional units and one memory port. Thus, TIC6x DSP processor may issue up to eight instructions including two memory fetches at the same time [20].

In the following sections, we first introduce our observation on selected three loops from the functions of EEMBC Telecommunication benchmark [10] and five loops from SMV benchmark [19] and their optimized assembly code generated by the TI C64 compiler. The algorithms used for de-pipelining and rerolling are presented in Sect. 3. A working example along with the experimental results is presented in Sects. 4 and 5. Sections 6 and 7 are related work and our summary.

2 Observation

We use data dependency graph (DDG) to represent a loop and follow graph theory to check whether or not a loop is re-rollable and if so, loop rerolling is performed. It is noted that if a loop is unrolled by a compiler, original DDG of the loop is always duplicated, resulting in an identical set of subgraphs referred to as subDDGs in this paper. To facilitate the discussion, we introduce some concepts below:

Two subDDGs G and H are said to be isomorphic if and only if the two subDDGs have the same node sets and any two nodes have a data dependence edge in G, their corresponding nodes in H have the same dependence edge. Isomorphism is an important concept in graph theory. If a DDG can be split into n isomorphic subDDGs, then the loop is re-rollable.

However, compilers often perform addition optimizations after loop unrolling which almost always cause changes to some subDDGs such that not all subDDGs are isomorphic. For example, TI compiler replaces single-word instructions with more efficient double-word instructions. It also uses peephole optimization to remove some instructions in some subDDGs. In fact, after analyzing the TI compiler optimized assembly code of the eight selected unrolled loops from SMV and EEMBC telecommunication benchmarks, we observed that not all subDDGs of the eight loops are isomorphic. In order to reroll the loop, all altered subDDGs must be converted back to isomorphic form.

To systematically tackle the complexity of the conversion process, we subdivide the loops into five different types.

0. A loop whose subDDGs are all isomorphic and all use the same index register and have the same operations on their corresponding nodes.
1. A loop that contains some memory fetch instructions using an additional index register for accessing the same array due to the limitation of instruction format when unrolling too many times.
2. A loop that uses two index registers to access the same array and an additional instruction in some subDDGs to move data across datapath, because memory fetch instruction must use the index register from its own datapath in the TI processor.
3. A loop that uses complex instructions of the TI processor to replace some simple instructions. For example, for performance enhancement a complex double-word LDDW instruction is used by the TI compiler to replace two single-word LDW instructions resulting in two subDDGs to share a single source node.
4. A loop with some of its instructions missing in its subDDGs due to peephole optimization.

Note that except for type_0, loops of all other types contain non-isomorphic subDDGs. As will be discussed in the following section, it is always possible to convert these non-isomorphic subDDGs back to isomorphic form. We name these non-isomorphic subDDGs isomorphicable in the following sections.

Figure 1 shows the categorization of subDDGs. Table 1 summarizes the characteristics of the eight unrolled loops selected as the target of the study in this paper. Table 2 lists subDDG features of the selected unrolled loops.

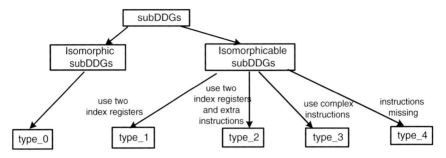

Fig. 1 Category of subDDGs

Table 1 Characteristics of unrolled loops

#	Function Name	Source code				Asm code				Loop optimization applied	
		Nest levels	Loop count			Nest levels	Loop count				
			outer	mid	inner		outer	mid	inner		
1	Dot product	1	-	-	100	1	-	-	50	unroll x2, then s/w pipelining	
2	Viterbi Decoder	1	-	-	31	0	-	-	0	unroll x31	
3	Viterbi StorePaths	1	-	-	32	1	-	-	7	unroll x4, then s/w pipelining	
4	SMV LSF_1	1	-	-	7	0	-	-	0	unrolling x7	
5	SMV LSF_2	3	9	128	10	2	9	128	0	innermost unrolling x10	
6	SMV LSF_3	3	9	128	7	2	9	128	0	innermost unrolling x7	
7	SMV LSF_4	2	7	-	10	1	7	-	0	inner unrolling x10 then outer s/w pipelining	
8	SMV FLT	2	170	-	9	1	85	-	0	1.inner unrolling x10 2.outer unrolling x2 3.s/w pipelining	outer_0 outer_1

3 Methodologies and Algorithms

Our methodologies for solving loop rerolling and software de-pipelining are described below:

1. Perform software de-pipelining first, then perform rerolling if the loop has been software pipelined after unrolling.
2. Build data dependency graphs of subDDGs based on the analysis of innermost loops in assembly code. The process begins from the *last_instructions* [18] to help reduce the search space.
3. Find clusters of potential unrolled copies including all isomorphic subDDGs and isomorphicable subDDGs.
4. Convert all isomorphicable subDDGs to isomorphic subDDGs using symbolic calculation, instruction replacing, de-peephole optimization and other techniques.
5. Use single loop to represent all isomorphic subDDGs, which is the rerolled loop.

Table 2 SubDDG features and de-optimization solution

#	Function Name	Loop Optimization Applied	Sub DDGs numbers	type	Features of Isomorphicable subDDGs	Solution of Loop De-optimization
1	Dot product	unroll x2, then s/w pipelining	2	0		software de-pipelining
2	Viterbi Decoder	unroll x31	31	1	two index registers	symbolic calculation
3	Viterbi StorePaths	unroll x4, then s/w pipelining	4	2	one extra MV instruction and two index registers	1. software de-pipelining 2. subDDG adjustment 3. symbolic calculation
4	SMV LSF_1	unrolling x7	7	3	use complex instructions	use simple instructions to replace complex instruction
5	SMV LSF_2	innermost unrolling x10	10	3	use complex instructions	use simple instructions to replace complex instruction
6	SMV LSF_3	innermost unrolling x7	7	4	one less instruction due to peephole optimization	de-peephole optimization
7	SMV LSF_4	inner unrolling x10 outer s/w pipelining	10	2	one extra MV instruction and two index registers	1. software de-pipelining 2. subDDG adjustment 3. symbolic calculation
8	SMV FLT	1. inner unrolling x10 _outer_0_	10	3	use complex instructions	1. software de-pipelining 2. use simple instructions to replace complex instruction
		2. outer unrolling x2 3. s/w pipelining _outer_1_	10	4	1. use complex instructions 2. two subDDGs have no load instruction due to peephole optimization 3. some subDDGs have extra MV instructions	1. software de-pipelining 2. use simple instructions to replace complex instruction 3. de-peephole optimization 4. subDDG adjustment

Figure 2 shows the flowchart of our loop de-optimization technique. Besides the normal control flow analysis and data flow analysis, we introduce the following 11 functions:

The **natural_loop_analysis** function:
From a given segment of assembly code and its control flow graph, the function finds the dominators, loop nest tree, loop headers, bodies, branches, nested loops, and the lengths of inner bodies. The algorithms of the function are very similar to that of [2].

The **software_pipelined_loop_checking** function:
The function checks all loops to find out whether the inner loop bodies are software-pipelined. If not, the execution of the software de-pipelining function is skipped.
    Algorithm:
    The algorithm checks for any pair of instructions op_i and op_j in the body of the inner loop and determines if the following conditions are true: (1) if op_i writes to a register which is to be read by op_j and op_j is located not earlier than op_i in the loop body and (2) if the latency of op_i is greater than the distance from op_i to op_j. If both of the conditions are true, then this loop has been software-pipelined because op_i and op_j cannot be in the same iteration.

The **software de-pipelining** function:
The function converts software pipelined loops to de-pipelined loop, the detailed description of the algorithm can be found in [5, 18].

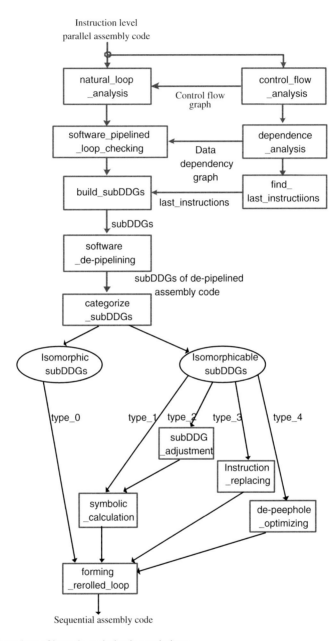

Fig. 2 Flow chart of loop de-optimization technique

The find_last_instructions function:
The function performs a bottom-up search of all de-pipelined loops for all last_instructions. A last_instruction belongs to either of the following two categories: (1) instructions that write to registers involving live variables with transferred values to be used after loop exits and (2) All memory store instructions.

The build_subDDGs function:
The function builds subDDGs for the bodies of all inner loops.
  Algorithm:

1. Set $subDDG_j = \{last\_instruction_j\}$ for each $last\_instruction_j$ in $subDDG_j$,
2. Define instruction $pool_j$ as the set of all instructions in de-pipelined loop body.
3. Add $instruction_k$ to $subDDG_j$ by performing a bottom up search for $instruction_k$ in instruction $pool_j$ from $last\_instruction_j$ where data precedes any instruction in $subDDG_j$ with true dependence, output dependence, or antidependence.
4. Repeat 3 until the first instruction in de-pipelined loop body has been reached.

The categorize_subDDGs function:
The function analyzes subDDGs and determines their types. It then selectively calls other functions depending on the type of the subDDG as described below.
  Algorithm:
  If subDDGs are isomorphic (i.e., type_0)
    {
    Use the same index registers and call forming_rerolled_loop function to
        reroll all isomorphic subDDGs;
    }
  Else subDDGs are isomorphicable
    {
    If type_1, call symbolic_calculation function;
    If type_2, call call subDDGs_adjustment and symbolic_calculation
            functions;
    If type_3, call instruction_replacing;
    If type_4, call de-peephole_optimization function and other functions;
    }

The symbolic_calculation function:
This function merges two different index registers. It does so by tracing back to the original source index register, replace it by a virtual register and recalculate all indexes.

The instruction_replacing function:
The function replaces a complex 32-bit instruction by two16-bit instructions with the same source and destination registers.

The subDDG_adjustment function:
The function applies to type_2 subDDG that uses a MV instruction to move data across datapath. This subDDG is semantically equivalent to the rest of subDDGs, therefore removing that MV instruction does not change the semantics.

The **de-peephole_optimizing** function:
The function recovers removed nodes and converts type_4 isomorphicable subDDG to isomorphic subDDG as some isomorphicable subDDGs have some of their nodes removed due to peephole optimization. For example, in one isomorphicable sub-DDG of SMV FLT a multiplication instruction node misses a load instruction node to provide its operand because peephole optimization removed this load instruction and the operand of that multiplication instruction is provided by another load instruction shared with another multiplication instruction. Another example is with SMV LSF_3 one isomorphicable subDDG in which one node of MV instruction is removed because the destination register of that MV instruction is dead.

Algorithm:
Compare a type_4 subDDG$_k$ with the isomorphic subDDG

1. If node$_i$ is found in isomorphicable subDDG$_k$ and its preceding node is missing in subDDG$_k$, then:

    i. Find node$_j$' that precedes node$_i$' in isomorphic subDDG where the corresponding node$_j$ in subDDG$_k$ is missing.
    ii. Copy node$_j$' and attach it to isomorphicable subDDG$_k$ such that the attached node precedes node$_i$.

2. If node$_i$ is found in isomorphicable subDDG$_k$ with its succeeding node missing, then:

    i. Find node$_j$' that succeeds node$_i$' in isomorphic subDDG but node$_j$ is missing in isomorphicable subDDG$_k$ as a succeeding node to node$_i$.
    ii. Make a copy of node$_j$' and add it to isomorphicable subDDG$_k$ as a succeeding node to node$_i$. If the destination register of node$_j$ is dead in isomorphicable subDDG$_k$, then convert isomorphicable subDDG$_k$ to isomorphic subDDG.

The **forming_rerolled_loop** function:
The function performs the following operations:

1. Replace all isomorphic subDDGs by a single subDDG.
2. Use list scheduling from last_instructions to arrange the partial order list of this subDDG in a bottom-up manner.
3. Add a backward branch instruction to form the rerolled loop body if no branch instruction in found in this subDDG.
4. Adjust loop count

4 Working Example

We have selected the StorePaths function in Viterbi of EEMBC telecommunication benchmark as a working example to demonstrate our loop rerolling and de-software pipelining techniques.

Figure 3a is its assembly code generated by TI C64 complier where each line is an instruction group and all instructions in one instruction group are executed at the same time in parallel.

(a)

```
1   $C$21:   LDW *+DP(_BufSelector),B4
2            MVKL   _BufPtr,B5
3            MVKH  _BufPtr,B5
4            MVK   7,A0
5            MVK   0x1,A1
6            LDW   *+B5[B4],B4
7            SUB  A4,8,B7
8            NOP     3
9            SUB  B4,16,B5
10           NOP    1
11           MV   B5,A3
12  $C$L4:   ; PIPED LOOP PROLOG
13  $C$L5:   ; PIPED LOOP KERNEL
14           NOP      3
15            SHR  A5,5,A4
16           [!A1] STH  A4,*++B7(8)
17           [!A1]  LDH   *+A3(4),B6
18           NOP      1
19           EXTU  A5,27,27,A4
20           NOP  1
21           MV   A4,B4
22           [!A1]  STH  B4,*++B5(16)      SHR  B6,5,B4
23           [!A1]  STH  B4,*+B7(2)
24           [!A1]  LDH   *+A3(8),A5
25           NOP      3
26           EXTU  B6,27,27,B4
27           [!A1]  STH  B4,*+B5(4)        SHR  A5,5,B4
28           [!A1]  STH  B4,*+B7(4)
29           [!A1]  LDH   *+A3(12),A4
30           NOP      2
31           [ A0]  BDEC $C$L5,A0
32           NOP      1
33           EXTU  A5,27,27,A5            SHR  A4,5,B4
34           [!A1]  STH  A5,*+A3(8)       EXTU A4,27,27,A4    [!A1] STH B4,*+B7(6)
35           LDH  *+A3(16),A5
36           [ A1]  SUB  A1,1,A1          [!A1]STH A4,*-A3(4)
37  $C$L6:   ; PIPED LOOP EPILOG
38           NOP      3
39           SHR  A5,5,A4
40           STH  A4,*++B7(8)
41           LDH   *+A3(4),B6
42           EXTU  A5,27,27,A4
43           NOP      2
44           MV  A4,B4
45           STH  A4,*++B5(16)            SHR B6,5,B4
46           STH B4,*+B7(2)               EXTUB6,27,27,B4
47           STH B4 *+B5(4)               LDH  *+A3(8),A4
48           NOP      4
49           SHR A4,5,B4
50           STH  B4,*+B7(4)
51           LDH   *+A3(12),A5
52           EXTU A4,27,27,A4
53           RETNOP B3,2
54           SHR  A5,5,B4
55           STH  A4,*+A3(8)              STH  B4,*+B7(6)     EXTU A5,27,27,A5
56           STH    A5,*+A3(12)
57           ; BRANCH OCCURS
```

(b)

```
1   $C$DW$21   LDW *+DP(_BufSelector),B4
2              MVKL _BufPtr,B5
3              MVKH _BufPtr,B5
4              MVK  8,A0
5              MVK  0x1,A1
6              LDW  *+B5[B4],B4
7              SUB  A4,8,B7
8              NOP    3
9              SUB  B4,16,B5
10             NOP    1
11             MV   B5,A3
12  $C$L5:     ; de-PIPED LOOP KERNEL
13             LDH *++A3(16),A5
14             NOP      4
15             SHR  A5,5,A4
16             STH  A4,*++B7(8)
17             LDH  *+A3(4),B6
18             NOP      1
19             EXTU  A5,27,27,A4
20             NOP 1
21             MV  A4,B4
22             STH  B4,*++B5(16) SHR B6,5,B4
23             STH  B4,*+B7(2)
24             LDH  *+A3(8),A5
25             NOP    3
26             EXTU  B6,27,27,B4
27             STH B4,*+B5(4)    SHR A5,5,B4
28             STH   B4,*+B7(4)
29             LDH   *+A3(12),A4
30             NOP     2
31             [A0] BDEC $C$L5,A0
32             NOP      1
33             EXTU  A5,27,27,A5 SHR  A4,5,B4
34             STH A5,*+A3(8)    EXTU A4,27,27,A4 STH B4,*+B7(6)
35             NOP      1
36             STH A4,*A3(4)
```

Fig. 3 Working example. **a** Assembly code of Viterbi StorePaths. **b** After software de-pipelining. **c** SubDDGs. **d** subDDGs adjustment. **e** Indexes of memory load and store instructions. **f** Rerolled loop of Viterbi StorePaths

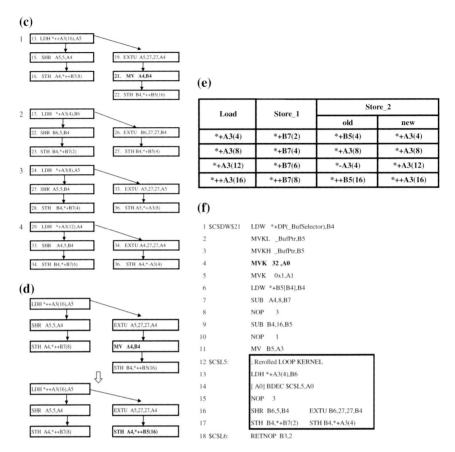

Fig. 3 (continued)

The iteration number of this loop body is seven. By using software_pipelined_loop_checking function, it is determined that this loop is software pipelined because register A5 is written by instruction LDH *++A3(16), A5 at line 35 and register A5 is read by instructions SHR A5,5,A4 and EXTU A5,27,27,A4 at lines 15 and 19, respectively; both instructions occur earlier than instruction LDH *++A3(16),A5.

Figure 3b shows the result of the software_de-pipelining function where the iteration number of de-pipelined loop body changes to eight. There are eight STH store instructions as last_instructions found by the find_last_instructions function.

Figure 3c is the result generated by build_subDDGs function. From the categorize_subDDGs function, we find that Viterbi StorePaths has four unrolled loop copies of type_2. The instruction numbers in Fig. 3c tie to the line numbers of instructions in Fig. 3b.

Among the four loop copies, one isomorphicable subDDG has one additional MV instruction generated by TI compiler for the purpose of moving data to another datapath. Figure 3d shows the semantically equivalent subDDGs before and after the removal of the MV instruction by the subDDGs_adjustment function.

After the above operations, we now have four type_1 subDDGs that are not yet isomorphic. This is because there are one load instruction and two store instructions in each unrolled copy, and the second store instructions of the four unrolled copies use different index registers. After the execution of symbolic_calculation function, all unrolled copies use the same index register for the second store instruction. Figure 3e lists the indexes of all memory load and store instructions, indicating that all subDDGs are now isomorphic and thus rerollable.

Figure 3f is the rerolled loop after the execution of forming_rerolled_loop function, which is semantically equivalent to the original assembly code shown in Fig. 3a. The iteration number of rerolled loop body changes to 32. The comparison before and after loop de-optimization is shown in Table 3, which is discussed in more detail in Sect. 5.

5 Experiment

We have chosen eight loop examples to conduct experiments manually. The original sets of assembly code are generated by TIC64 compiler, which are then optimized by loop unrolling and/or software pipelining. Their characteristics are summarized in Table 1. Their subDDG features and the solutions of de-optimization are summarized in Table 2.

Besides Dot product, Viterbi Decoder and Viterbi StorePaths are from Viterbi function of EEMBC Telecommunication benchmark. The other five kernels are from LSF_Q_New_ML_search_fx and FLT_filterAP_fx functions of the SMV benchmark. Table 1 presents the number of nested levels and loop counts of the source code and assembly code; it also shows the optimization methods applied by TI C64 compiler. All examples have loop unrolling; some involve both loop unrolling and software pipelining. In addition, Table 2 presents the characteristics of subDDGs, the types of isomorphicable subDDGs, the causes for their occurrences, as well as the solution for loop de-optimization. Dot product is the simplest example; all its subDDGs are isomorphic subDDGs using the same index register. The function categorize_subDDGs determines it is type_0 and the forming_rerolled_loop function can thus be called immediately. The remaining examples need conversion from isomorphicable subDDGs to isomorphic subDDGs. SMV FLT is the most complicated case, in which the compiler unrolls the inner loop first, and then unrolls the code of outer loops, and finally software pipelines them. Moreover, peephole optimization is used to reduce some instructions, which further complicates the rerolling process. In general, loop de-optimization requires a range of activities and techniques including software_de-pipelining, instruction_replacing, subDDG_adjustment, de-peephole_optimizing, and finally forming_loop_rerolling.

Table 3 presents our experimental results, where #I denotes number of instructions; #IG number of instruction groups; #CC clock cycles which represents the execution

Table 3 Experimental results

#	Function name	Original				After de-pipelining				After rerolling			
		#I	#IG	#CC	#LC	#I	#IG	#CC	#LC	#I	#IG	#CC	#LC
1	Dot product	44	19	113	50	14	12	552	50	11	8	802	100
2	Viterbi Find-Metrics	50	38	38	0	–	–	–	–	21	15	204	32
3	Viterbi Store-Paths	65	4	208	7	40	35	211	8	8	6	243	32
4	SMV LSF_1	13	3	13	0	–	–	–	–	6	6	37	7
5	SMV LSF_2	86	3	40	0	–	–	–	–	13	11	120	10
6	SMV LSF_3	37	9	19	0	–	–	–	–	21	11	48	6
7	SMV LSF_4	144	5	125	7	71	38	146	7	10	8	630	70
8	SMV FLT	237	92	2997	85	171	60	5100	85	29	38	13260	outer 170 inner 9

time of specific code used in the experiment; and #LC loop count. There are three sections in Table 3, the leftmost one is original assembly code, and the rightmost section is the final result of the semantically equivalent sequential code after loop de-optimization. The second section lists certain kernels that have been optimized by software pipelining after loop unrolling by the compiler. Based on the final results of loop de-optimization, it is obvious that code sizes, including both instruction count and number of instruction groups, are reduced while the number of clock cycles is increased.

6 Related Work

Since Cifuences and her colleagues presented their work [9], many decompilation techniques have been published [1, 8, 12]. However, few papers tackle deoptimizaton technique and fewer still investigate loops with instruction-level parallel architectures.

Snavely et al. [16] present instruction level deoptimization approaches on Intel Itanium including unpredication, unscheduling and unspeculation. However they did not tackle loop de-optimization and software de-pipelining. Wang et al. [21] apply un-speculation technique on modulo scheduled loops to make the code easier to understand, however they do not tackle software de-pipelining and loop rerolling.

Loop rerolling has been implemented at source code level in LLVM compiler [15]. Stitt and Vahid [17] use loop rerolling technique for binary-level coprocessor generation, which is the reverse of loop unrolling. They use character string to represent instruction sequence, and then use the suffix tree to represent the character string for efficient pattern matching unrolled loop copies. However, their techniques are applicable only to decompiling assembly code such as MIPS and ARM without instruction level parallelism.

Bermudo et al. present an algorithm for reconstruction of the control flow graph for assembly language program with delayed instructions which was used in a reverse compiler for TI DSP processors [4]. Su et al. present software de-pipelined technique [18] for single-level loops. Their method based on building linear data dependency graph in software pipelined loop can convert the complicated software pipelined loop code to a semantically equivalent sequential loop. Bermudo et al. extend software de-pipelined technique to nested loops [5].

7 Summary

We present our instruction level loop de-optimization algorithms which involve software de-pipelining and loop rerolling. Instruction level loop de-optimization can be very complicated, particularly when the assembly code after loop unrolling is combined with peephole optimization. It is noted that although different compilers may generate different optimized assembly code, our approach can be a useful technique to help interested readers gain insight especially in the difficult tasks of loop rerolling and software de-pipelining, the necessary steps to decompile loops at instruction level. In this paper, we consider only loop independent dependency and plan to extend it to handle loop carried dependency in the future.

Acknowledgments Su would like to thank the ART awards of William Paterson University.

References

1. Anand, K., et al.: Decompilation to compiler high IR in a binary rewriter. Technical report, University of Maryland (2010)
2. Aho, A., et al.: Compilers: Principles, Techniques, and Tools, 2nd edn. Addison-Wesley (2007)
3. Bermudo, N., et al.: Control flow graph reconstruction for reverse compilation of assembly language programs with delayed instructions. In: Proceedings of SCAM2005, pp. 107–118 (2005)
4. Bermudo, N.: Low-level reverse compilation techniques. Ph.D. thesis. Technische Universität Wien (2005)
5. Bermudo, N., et al.: Software de-pipelining for nested loops. In: Proceedings of IMCEEME'12, pp. 39–44 (2012)

6. Cesare, S.: Detecting bugs using decompilation and data flow analysis. In: Black Hat USA 2013, https://media.blackhat.com/.../US-13-Cesare-Bugalyze.com (Accessed 2013)
7. Cesare, S., et al.: Malwise—an effective and efficient classification system for packed and polymorphic malware. IEEE Trans. Comput. 1193–1206 (2013)
8. Chen, G., et al.: A refined decompiler to generate C code with high readability. In: Proceedings of the 17th Working Conference on Reverse Engineering (2010)
9. Cifuentes, C.: Reverse compilation techniques. Ph.D. Dissertation, Queensland University of Technology, Department of CS (1994)
10. Hu, E., et al.: New DSP benchmark based on selectable mode vocoder (SMV). In: Proceedings of the 2006 International Conference on Computer Design, pp. 175–181 (2006)
11. Křoustek, J.: Decompilation of VLIW executable files—caveats and pitfalls. In: Proceedings of Theoretical and Applied Aspects of Cybernetics (TAAC'13), pp. 287–296. TSNUK, Kyiv (2013)
12. Křoustek, J., Kolář, D.: Preprocessing of binary executable files towards retargetable decompilation. In: Proceedings of the 8th International Multi-Conference on Computing in the Global Information Technology (ICCGI'13), pp. 259–264. IARIA, Nice (2013)
13. Lam, M.: Software pipelining: an effective scheduling technique for VLIW machines. In: Proceedings of the SIGPLAN 88 Conference on PLDI, pp. 318–328 (1988)
14. Lavery, L., Hwu, H.: Unrolling based optimizations for modulo scheduling. Proc. MICRO **28**, 328–337 (1995)
15. LLVM: LoopRerollPass.cpp Source File. http://llvm.org/docs/doxygen/html/LoopRerollPass_8cpp.html (Accessed 2014)
16. Snavely, N., Debray, S.: Unpredication, unscheduling, unspeculation: reverse engineering Itanium executables. IEEE Trans. Softw. Eng. **31**(2) (2005)
17. Stitt, G., Vahid, F.: New decompilation techniques for binary-level co-processor generation. In: Proceedings of the International Conference on Computer Aided Design—ICCAD, pp. 547–554 (2005)
18. Su, B., et al.: Software de-pipelining technique. In: Proceedings of the SCAM2004, pp. 7–16 (2004)
19. TELE BENCH, an EEMBC Bench, http://www.eembc.org/benchmark/telecom_sl.php (Accessed 2015)
20. TMS320C64x/C64x+ DSP CPU and Instruction Set Reference Guide, SPRU732H (2008)
21. Wang, M., et al.: Un-speculation in modulo scheduled loops. In: Proceedings of the 2nd International Multisymposium on Computer and Computational Sciences, pp. 486–489 (2008)

ETL Design Toward Social Network Opinion Analysis

Afef Walha, Faiza Ghozzi and Faïez Gargouri

Abstract Now-a-days, social networking sites have been created lot of buzz in technology world. They are considered as a rich source of information because people share and discuss their opinions about a certain topic freely. Sentiment analysis or opinion mining is used for knowing voice or response of crowd for products, services, organizations, individuals, events, etc. Due to the importance of user's opinions in decisional systems, several Data Warehouse approaches integrate them through a cleaning and transformation processes. However, there is a clear lack of a standard model that can be used to represent the ETL processes. We propose an ETL design approach integrating user's opinion analysis, expressed on the popular social network Facebook. It consists in the extraction of opinion data on Facebook pages (e.g. comments), its pre-processing, sentiment analysis and classification, reformatting and loading into the Data WeBhouse (DWB).

Keywords ETL · Opinion analysis · Social network

1 Introduction

The Web has dramatically changed the way that people express their views and opinions. They can now post reviews of products at merchant sites and express their views on almost anything in Internet forums, and social networking sites (e.g. Facebook, twitter), which are collectively called user-generated content. This online behavior represents new and measurable sources of information to an organization. For a company, it may no longer be necessary to conduct surveys, organize focus

A. Walha (✉)
Multimedia, Information systems and Advanced Computing Laboratory,
MIR@CL, University of Sfax, Sfax, Tunisia
e-mail: afef_walha@yahoo.fr

F. Ghozzi · F. Gargouri
Institute of Computer Science and Multimedia, University of Sfax, Sfax, Tunisia
e-mail: jedidi.faiza@gmail.com

F. Gargouri
e-mail: faiez.gargouri@gmail.com

© Springer International Publishing Switzerland 2016
R. Lee (ed.), *Computer and Information Science 2015*,
Studies in Computational Intelligence 614, DOI 10.1007/978-3-319-23467-0_16

235

groups or employ external consultants in order to find consumer opinions about its products and those of its competitors because the user-generated content on the Web can already give them such information.

With the growing popularity of social networks, millions of users interact frequently and share variety of digital content with each other. They express their feelings and opinions on every topic of interest. These opinions carry import value for personal, academic and commercial applications. Social network sites contain a lot of customers' opinions on certain products that are helpful for decision making. In spite of this importance, there is a clear lack of a standard model that can be used to represent the ETL processes (Extraction, Transformation and Loading) of social networking sites integrating opinion data. In this paper, we propose to design ETL processes for Facebook page interactions.

This paper is organized as follow: Sect. 2 presents a brief review on ETL design and sentiment analysis approaches. Then, we detail our proposed ETL design processes applied on Facebook pages. Finally, we conclude and present some perspectives in Sect. 4.

2 Background

This section deals with two main aspects in the literature: ETL design processes and opinion analysis methods and techniques.

2.1 ETL Modeling Approaches

ETL processes design is a crucial task in DW development due to its complexity and its time consuming. Works dealing with this task [2, 3, 12, 14, 16, 17] can be classified into two main groups: Specific ETL modeling and Standard ETL modeling. The first group [3, 16] offers specific notations and concepts to give rise for new specialized modeling languages. Extraction, transformation and loading processes proposed in [16] are limited to typical activities (e.g. join, filter). El-Sappagh et al. [3] extends these proposals by modeling advanced operations, like user define functions and conversion into structure, etc. In order to design complex ETL scenario, specific modeling approaches propose conceptual and formal models. However, the standardization is an essential asset in modeling. The goal of the second group is to overcome this problem by using modeling languages such as UML and BPMN. Trujillo and Luján-Mora [14] and Muñoz et al. [12] use UML class diagram to represent ETL processes statically or dynamically by using UML activity diagram. Wilkinson et al. [17] and Akkaoui et al. [2] use BPMN standard where ETL processes can be a particular type of business.

Even though ETL processes modeling approaches succeeded in providing interesting several modeling languages, they don't cover opinion data sources available on Web resources like social networks, blogs, reviews, etc.

2.2 Opinion Analysis Approaches

Opinions are usually subjective expressions that describe people's sentiments, appraisals or feelings toward entities, events and their properties. The concept of opinion is very broad. In this paper, we focus only on opinion expressions that convey people's positive and negative sentiment.

Integrating opinion data is nowadays a hot topic for many researchers. The common goal of opinion analysis approaches is to detect text polarity: positive, negative or neutral. In Medhat et al. [9], categorize sentiment analysis approaches into machine learning and lexicon approaches. Machine learning approaches [1, 18] use classification techniques to classify text (e.g. Naive Bayes (NB), maximum entropy (ME), and Support Vector Machines (SVM)). Lexicon approaches [5, 6, 8, 11, 13] rely on a sentiment lexicon, a collection of known and precompiled sentiment terms. They use sentiment dictionaries with opinion words and match them with the data to determine text polarity. They assign sentiment scores to opinion words according to positive or negative words contained in the dictionary. Lexicon-based approaches are divided into dictionary-based approaches and corpus-based approaches.

A Dictionary-based approach [7, 11] begins with a predefined dictionary of positive and negative words, and then uses word counts or other measures of word incidence and frequency to score all the opinions in the data. The idea of these approaches is to firstly collect manually a small set of opinion words with known orientations (seed list), and then to grow this set by searching in a known lexical DB (e.g. WordNet dictionary) for their synonyms and antonyms. The newly found words are added to the seed list [8]. Opinion words share the same orientation as their synonyms and opposite orientations as their antonyms. In [5, 13], authors use this technique to find semantic orientation for adjectives. Qiu et al. [13] worked on web forums to identify sentiment sentences in contextual advertising. They used syntactic parsing and sentiment dictionary and proposed a rule-based approach to tackle topic word extraction and consumers' attitude identification in advertising keyword extraction.

Corpus based techniques rely on syntactic patterns in large corpora. Corpus-based method can produce opinion words with relatively high accuracy. A corpus-based method needs very large labeled training data. Jiao and Zhou [6] used the Conditional Random Fields (CRFs) methods in order to discriminate sentiment polarity by multi-string pattern matching algorithm applied on Chinese online reviews in order to identify sentiment polarity. They established emotional and opinion words dictionaries.

Machine learning and lexicon approaches use opinion words and classification techniques to determine text polarity. In addition to the use of opinion words to analyze sentiments, emoticons decorating a text can give a correct insight of the sentence or text. For example, the emoticon "☺" expressing "happiness" means positive opinion. Further researchers take care of the increasing using of these typographical symbols for sentiment classification [4, 15]. Vashisht and Thakur [15] identify the possible set of emoticons majorly used by people on Facebook and use them to classify text polarity. Then, they used a finite state machine to find out the polarity of the sentence or paragraph. The problem with this approach is performing sentiment analysis on text-based status updates and comments, disregarding all verbal information and using only emoticons to detect both positive and negative opinions. Hogenboom et al. [4] propose a framework for automated sentiment analysis, which takes into account information conveyed by emoticons. The goal of this framework is to detect emoticons, determine their sentiment, and assign the associated sentiment to the affected text in order to correctly classify the polarity of natural language text as either positive or negative.

Existing ETL design approaches model various web sources without considering user's opinions available on social networks, reviews, blogs, forums or emails, etc. In the past few years, many researchers have shown interest to opinions expressed by people on any topic. They proposed sentiment analysis methods and techniques to determine text polarity. Some approaches apply classification algorithms and use linguistic features (machine learning approaches). Others use sentiment dictionaries with opinion words and match them with data sources to determine polarity (lexicon-based approaches). These approaches assign sentiment scores to opinion words according to positive or negative words contained in the dictionary. Others researchers use emoticons to disambiguate sentiment when sentiment is not conveyed by any clearly positive or negative words in a text segment.

Sentiment analysis approaches presented in the literature are very helpful and interesting in order to classify a text (positive or negative polarity). In spite of the importance of sentiment classification approaches, we note that few of them employ the coupling between opinion analysis and ETL processes in order to enhance semantic orientation to multidimensional design.

In the current work, we define a new approach of ETL processes design integrating people's opinions exchanged on Facebook social network. Facebook users express their opinions about any topic freely through opinion words and emoticons. Sentiment analysis is required to classify user opinion. For that, we adopt a lexicon approach based on dictionaries used as lexical DBs in our ETL processes design. We are based on the modeling standard BPMN 2.0 to design Extraction, Transformation and Loading processes because of its completed graphical notation in modeling business processes understandable by all business categories of users [2].

3 Proposed ETL Processes Modeling

DWB (Data WeBhouse) sources may include several data types, such as geographic DBs, web sites DBs, web logs, language recognition systems and social networking sites, etc. In order to enrich ETL processes design with semantic orientations, we are interested to opinion data shared and discussed freely on the popular social network Facebook.

Our ETL design approach provides to company's ETL designers a framework integrating costumers' opinions about their products or services. User actions (comments, messages, posts and likes) exchanged within Facebook pages are pertinent for marketing and advertising industry to gather opinions about a particular product. The goal of our ETL design approach is to analyze user actions on product features in order to classify his opinion (positive or negative). To assume this analysis, we are based on verbal cues: opinion words and graphical cues: emoticons. For that, we identify two dictionaries: opinion dictionary composed of opinion words (e.g. best, good) and emoticons dictionary (e.g. :), :(, ;), etc.).

3.1 Lexical DB Description

Opinion and emoticons dictionaries serve as lexical DB in our ETL design approach. Opinion dictionary is composed of opinion words that express desirable (e.g. great, amazing, etc.) or undesirable (e.g. bad, poor, etc.) states. Emoticons dictionary contains positive (e.g. ☺) and negative (e.g. ☺) emoticons majorly used by Facebook users.

Figure 1 illustrates the process of defining our lexical DB. To identify opinion dictionary, we follow a dictionary-based method [8]. Its main idea is to manually collect a small set of terms (seed words), and then search in the well known corpora WordNet [10] of their synonyms and antonyms to enrich them. Then, a manual inspection is carried out to remove or correct errors existing in opinion dictionary. In some texts, opinion word can be related to a modifier term that changes its sentiment polarity (e.g. in the sentence "it is not beautiful", the modifier term "not" changes the sentiment polarity of the opinion word "beautiful"). Also, amplifier terms can increase or decrease the polarity of the affected opinion word (e.g. the word "very" in the sentence "it's very big" increase the polarity of the opinion word "big"). For that, we classify opinion words to two types: modifier terms (like "not" and "very") and carrying-sentiment terms (such as "big", "beautiful").

With the increasing use of emoticons, it is of utmost importance to consider these typographical symbols to discriminate sentiment polarity. So, we collect a set of emoticons majorly used by people on Facebook including positive and negative emoticons defined in [15].

The final step in lexical DB definition process (Fig. 1) is to associate polarity score to each opinion dictionary term and emoticon already defined in opinion and

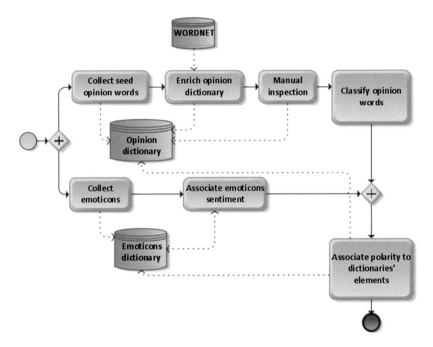

Fig. 1 Lexical DB definition process

emoticons dictionaries. This score has positive (between (0) and (1)) or negative (between (−1) and (0)) value. This real value is determined by linguistic experts according to their sentiment classification. The positive polarity (0.8) is then associated to the opinion word "enjoy" expressing "Happiness" sentiment. Tables 1 and 2 detail examples of carrying-sentiment words and modifiers defined in opinion word dictionary. Moreover, Table 3 shows examples of emoticons and their associated polarities.

Dictionaries defined in this process aims to determine the sentiment polarity of opinions expressed on product features in Facebook pages. Emoticons and opinion dictionaries are used in our ETL processes design to analyze user actions in order to be transformed to DWB model.

3.2 ETL Processes Design

Our ETL scenario aims to capture Facebook data through Facebook API graph explorer, bring it to an adapted format and feed the transformed data into the target DWB.

Figure 2 is an overview of the proposed ETL processes: Extraction, Transformation and Loading. These processes are based on the lexical DB (opinion and

Table 1 Examples of sentiment-carrying words, their associated sentiments and polarity scores

Sentiment-carrying word				Sentiment classification	Polarity score
Love	Lovely	Loved	Loving	Heart	0.9
Like					0.8
Hate		Dislike			−0.8
Curious					0.5
Fantastic		Perfect		Quality	1
Fabulous	Best		Excellent		0.9
Better					0.8
Good					0.7
Bad		Badly			−0.8
Worst					−0.9
Happy				Happiness	0.8
Excited					0.95
Unhappy				Sadness	−0.7
Depressed					−1
Amazing				Amused	0.8

Table 2 Examples of modifiers and their associated polarity scores

Modifier term			Modifier Polarity
So			1
Totally			0.9
Very			0.8
Small			0.5
Little		Few	0.4
Not	Don't	Didn't	−1

emoticons dictionaries) to analyze user actions expressed on products features within Facebook pages. The result of this analysis is to determine polarity score reflecting user's opinion.

3.3 Extraction Step

Extraction step is responsible for capturing data from different sources. According to DWB multidimensional schema (presented in Fig. 8), we aim to analyze user actions associated to posts shared on Facebook pages. A post is an individual entry of a user, page, or group. A list of available actions (comments and likes) is associated to each post. These actions can help to gather people's opinions related to a post.

Table 3 Examples of emoticons and their associated polarity scores

Emoticon			Sentiment classification	Polarity
:-)	=)	:]	Happiness/smile	0.6
:)				0.7
:-(			Sadness	–0.6
:(				–0.7
:'(				–0.9
:D			Amused	0.9
=D				0.8
<3			Love/heart	1
>:(			Anger	–0.64
(y)			Thumbs up	0.84

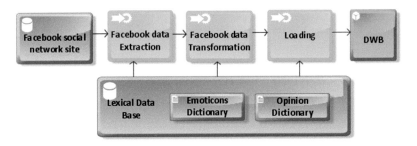

Fig. 2 ETL processes modeling framework

Figure 3 details extraction process. It starts by collecting general information about each Facebook page (page name, website, description, category, etc.). Then, it extracts posts shared on this page. The next step consists in extracting post information including source, message, picture, description, link, and created-time. Finally, this step collects actions (user likes and comments) associated to each post.

Figure 4 illustrates an example of post shared on "Sephora" Facebook page Comments associated to this post are shown in Fig. 5.

3.4 Transformation Step

Transformation step tends to make cleaning and conforming on DWB sources (Facebook page actions) to gain correct, complete, consistent, and unambiguous data.

Transformation step is organized in three main steps: pre-processing, analysis and mapping (see Fig. 6).

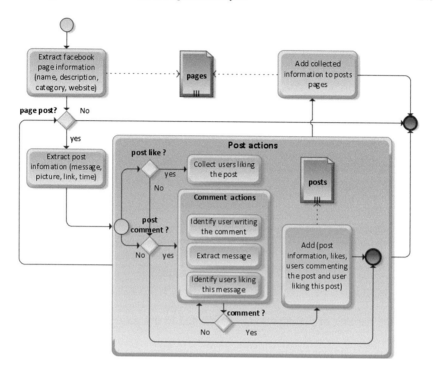

Fig. 3 Facebook page extraction process

message = "Milks nourish and hydrate beautifully—and are especially great during drying, cold weather. Are you a fan?"
created_time = "2015-01-29T02:00:01+0000"
type = "photo"
link = "https://www.facebook.com/Sephora/photos/..... /?type=1"

Fig. 4 Post (P) shared on the page "Sephora"

Pre-processing starts by comments cleaning which replaces all capital letters with small letters and removes diacritics. For example, in comment (1) (Fig. 5), the term "fabuloüs" is replaced by "fabulous". Then, it identifies each comment word POS (Part-Of-Speech) and its type, i.e., sentiment-carrying or modifying terms [16]. The latter change the sentiment of corresponding opinion word(s) such as negations that change the sentiment sign (e.g. the modifier "not", used in comment (6), change the sentiment polarity of the opinion term "good"). Also, amplifiers increase the sentiment of the affected sentiment words (e.g. the amplifier "very" in comment (2) modifies the sentiment of the opinion word "good").

Analysis is the main step of transformation process. It aims to calculate sentiment score of a post (P), i.e., Sent (P_U). This score is equal to the average of comments' sentiment scores associated to the post (P), as in (1), i.e.

$U_1 \Rightarrow C_1$: "I shop Sephora frequently!! Love !!! Love !!! Love !!!!"

$U_2 \Rightarrow C_2$: "Fabuloüs stuff!! fantastic, I love it."

$U_3 \Rightarrow C_3$: "I'm curious to try it"

$U_2 \Rightarrow C_4$: "Can't wait to try it! It is very good <3"

$U_4 \Rightarrow C_5$: "I'm :(badly product. not good"

$U_5 \Rightarrow C_6$: "Excellent sephora!"

$U_6 \Rightarrow C_7$: "I LOVE SEPHORA. My husband surprised me with this today when I came home.
So excited! And it really smells"

$U_5 \Rightarrow C_8$: "(y) Amazing milk!"

$U_7 \Rightarrow C_9$: "I am using this product. I'm unsatisfied !!!"

$U_4 \Rightarrow C_{10}$: "I don't like this cosmetic product >:("

$U_3 \Rightarrow C_{11}$: "I bought It. perfect !!"

$U_2 \Rightarrow C_{12}$: "LOVELY !!!"

Fig. 5 Examples of comments associated to the post (P)

$$\text{Sent } (P_U) = \frac{\sum_{j=1}^{N} \text{Sent}(C_i)}{N} \tag{1}$$

With N the number of comments (C_i) associated to the post (P) published by the user (U).

To compute sentiment score of the comment message (C_i), we propose a lexicon-based method. Its goal is to associate sentiment score to each comment (Sent (C_i)). The principle of this method is the following: if the comment (C_i) contains opinion words and emoticons, Sent (C_i) is computed as the average of all emoticons' sentiment polarities (Sent (e_{ij})) and polarities of sentiment-carrying words (w_{ij}) and their modifiers (m_{ij}). Otherwise, if the comment (C_i) contains opinion words without visual cues (emoticons), Sent (C_i) is calculated as the average of sentiment-carrying words (w_{ij}) and their modifiers (m_{ij}) polarities (if any, Sent (m_{ij}) defaults to 0). The sentiment score equation of the ith comment (C_i) is then defined in (2), i.e.,

$$\text{Sent } (C_i) = \begin{cases} \dfrac{\sum_{j=1}^{v_i} \text{Sent } (e_{ij}) + \sum_{j=1}^{t_i} \dfrac{|\text{Sent } (m_{ij})| + \text{Sent } (w_{ij})}{2} \times S(\text{Sent } (m_{ij}))}{v_i + t_i} & \text{if } v_i > 0 \\[2em] \dfrac{\sum_{j=1}^{t_i} \dfrac{|\text{Sent } (m_{ij})| + \text{Sent } (w_{ij})}{2} \times S(\text{Sent}(m_{ij}))}{t_i} & \text{else,} \end{cases} \tag{2}$$

With v_i and t_i correspond respectively to the number of emoticons and the number of sentiment-carrying words used in the comment (C_i). $S(\text{Sent } (m_{ij}))$ depends on the polarity ($+/-$) of the modifier (m_{ij}) related to opinion word. We assign the value (1) to $S(\text{Sent } (m_{ij}))$ if the modifier polarity is positive (Sent (m_{ij})>0). Otherwise, if the (m_{ij}) has a negative polarity, $S(\text{Sent } (m_{ij}))$is equal to (−1).

Comment sentiment analysis process, described in Fig. 7, details steps to determine comment's sentiment score (Sent). This process starts by computing the number of opinion words (t_i) and emoticons (v_i) used in the comment (C_i), and initializing (Sent) to the value (0). For each opinion word (w_{ij}) used in (C_i), it searches the mod-

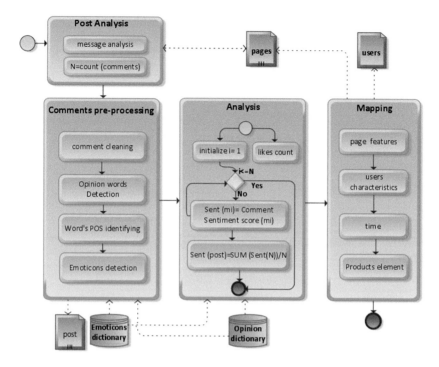

Fig. 6 Facebook page transformation process

ifier (m_{ij}) related to this word. If (m_{ij}) exists, it recuperates its sentiment score (Sent (m_{ij})) defined in opinion dictionary. The absolute value of this score (Sent (m_{ij})) is added to (w_{ij}) polarity score (Sent (w_{wj})), divided by (2), then multiplied by the modifier score polarity (S), i.e. (1) or (−1) and added to (Sent). Next, his process follows by determining sentiment polarity (Sent (e_{ij})) of each emoticon (e_{ij}) exploited in the comment (C_i) and add it to (Sent). The final step is to determine the final value of Sent, i.e. the average of opinion words and emoticons scores as defined in (2).

To determine users' opinions corresponding to the post (P), we apply "Comment sentiment score" process (Fig. 7) on a set of comments (Fig. 5) associated to this post (Fig. 4). Results are depicted in Table 4. 0Post's sentiment score (Sent (P_U)) is computed according to (1).

The final step in transformation process (Fig. 6) is the mapping. Its role is the matching between the source (concepts of "Facebook" model) and the target (DWB multidimensional elements). For example, the attribute "Category" of the class PAGE (source model) corresponds to the parameter "categoryPP" of the dimension FACE-BOOK POSTS (DWB multidimensional schema presented in Fig. 8).

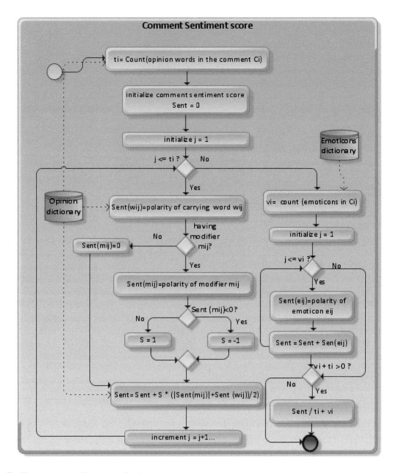

Fig. 7 Comment sentiment analysis process

Table 4 Comments' sentiment polarity

User U	U_1	U_2				U_3		U_4		U_5		U_6	U_7
Comment C_i	C_1	C_2	C_4	C_{12}	C_3	C_{11}	C_5	C_{10}	C_6	C_8	C_7	C_9
Sent (C_i)	0.9	0.93	0.87	0.9	0.5	1	−0.78	−0.77	0.9	0.8	0.93	−0.75
Sent (P_U)	0.9	0.9			0.75		−0.77		0.85		0.93	−0.75

3.5 Loading Step

The goal of loading process is to feed the DWB with data resulted from transformation step. It consists in loading data into DWB multidimensional elements including dimensions, measures, facts, attributes and parameters. These elements are illustrated in Fig. 8. The fact POST_ACTION analyzes user actions (comments and likes)

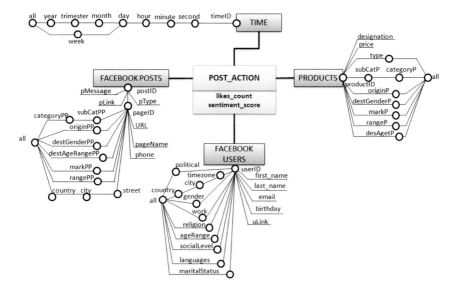

Fig. 8 DWB star schema

associated to a post commented by users on Facebook pages. Decisional makers can then analyze likes_count and sentiment_score according to TIME, PRODUCT, FACEBOOK POSTS and FACEBOOK USERS dimensions. For examples, Fig. 9 provides them with sentiment scores resulted from the analysis of users' comments associated to the post (P) corresponding to the product described in (P) on "Valentine's" day. Manager can notice that users (U_4) and (U_7) have negative opinions. So, he can define user profile interested to this product. Figure 10 shows also analysis results of comments shared by the user U_2 related to four products presented respectively in posts (P1), (P2), (P3) and (P4) during "February".

Fig. 9 Sentiment Polarity
Scores associated to
the post (P)

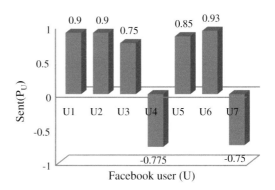

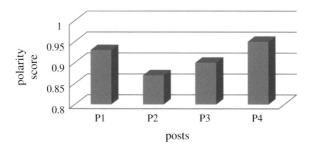

Fig. 10 Sentiment Analysis of the user "U2" on "February"

4 Conclusion and Future Works

Opinions are usually subjective expressions that describe people sentiments and appraisals. Social networks are platforms where millions of users interact frequently and express opinions on every topic of interest. Due to the importance of user opinions to decisional systems, we worked on integrating them DWB design.

We present in this paper a new ETL processes modeling approach using BPMN standard. This approach integrates user opinions expressed by comments shared on the social network Facebook. Its goal is to detect both positive and negative comment polarity. We associate for that a sentiment score depending on comments opinion terms and emoticons. This sentiment analysis is a lexicon method. This analysis is based on opinion and emoticon dictionaries to classify comment polarity.

As future works, we will evaluate our sentiment analysis process on a large test collection of user actions and enrich our lexical DB in order to adapt context-specific opinion analysis. Also, we will extend our ETL processes design approach by integrating more opinion web sources available on web logs, web sites and other social networks.

References

1. Abbasi, A., Chen, H., Salem, A.: Sentiment analysis in multiple languages: feature selection for opinion classification in web forums. ACM Trans. Inf. Syst. **26**(3), 1–34 (2008)
2. Akkaoui, Z.E., Mazón, J., Vaisman, A.A., Zimányi, E.: BPMN-Based conceptual modeling of ETL processes. In: 14th International Conference on Data Warehousing and Knowledge Discovery (DaWaK), pp. 1–14 (2012)
3. El-Sappagh, S., Hendawi, H., Bastawissy, A.H.: A proposed model for data warehouse ETL processes. J. King Saud Univ. Comput. Inf. Sci. **23**(2), 91–104 (2011)
4. Hogenboom, A., Bal, D., Frasincar, F.: Exploiting emoticons in sentiment analysis. In: Proceedings of the 28th Annual ACM Symposium on Applied Computing (SAC), pp. 703–710 (2013)
5. Hu, Y., Li, W.: Document sentiment classification by exploring description model of topical terms. Comput. Speech Lang. J. **25**(2), 386–403 (2011)

6. Jiao, J., Zhou, Y.: Sentiment polarity analysis based multi dictionary. In: International Conference on Physics Science and Technology (ICPST) (2011)
7. Kim, S., Hovy, E.: Determining the sentiment of opinions. In: Proceedings of the 20th International Conference on Computational Linguistics (COLING'04) (2004)
8. Liu, B.: Web Data Mining: Exploring Hyperlinks, Contents, and Usage Data, 2nd edn. Springer, New York (2011)
9. Medhat, W., Hassan, A., Korashy, H.: Sentiment analysis algorithms and applications: a survey. Ain Shams Eng. J. **5**(4), 1093–1113 (2014)
10. Miller, G., Beckwith, R., Fellbaum, C., Gross, D., Miller, K.: WordNet: an on line lexical database. In: International Journal of Lexicography, vol. 3. Oxford University Press (1990)
11. Minging, H., Bing, L.: Mining and summarizing customer reviews. In: Proceeding of ACM SIGKDD International Conference on Knowledge Discovery and Data Mining (KDD'04), pp. 168–177 (2004)
12. Muñoz, L., Mazón, J.N., Trujillo, J.: A family of experiments to validate measures for UML activity diagrams of ETL processes in data warehouse. Inf. Softw. Technol. **52**(11), 1188–1203 (2010)
13. Qiu, G., He, X., Zhang, F., Shi, Y., Bu, J., Chen, C.: DASA: dissatisfaction-oriented advertising based on sentiment analysis. Expert Syst. Appl. J. **37**(9), 6182–6191 (2010)
14. Trujillo, J., Luján-Mora, S.: A UML based approach for modeling ETL processes in data warehouses. In: 22nd International Conference on Conceptual Modeling—ER. Lecture Notes in Computer Science, vol. 2813, pp. 307–320 (2003)
15. Vashisht, S., Thakur, S.: Facebook as a corpus for emoticons-based sentiment analysis. Int. J. Emerg. Technol. Adv. Eng. J. (IJETAE) **4**(5), 904–908 (2014)
16. Vassiliadis, P.: A survey of extract–transform–load technology. Int. J. Data Wareh. Min. (IJDWM), **5**(3), 1–27 (2009)
17. Wilkinson, K., Simitsis, A., Dayal, U., Castellanos, M.: Leveraging business process models for ETL design. In: ER 2010: 29th International Conference on Conceptual Modeling, November 2010. Lecture Notes in Computer Science, vol. 6412, pp. 15–30 (2010)
18. Wilson, T., Wiebe, J., Hoffmann, P.: Recognizing contextual polarity in phrase-level sentiment analysis. In: Proceeding of the conference on Human Language Technology and Empirical Methods in Natural Language Processing. Association for Computational Linguistics, pp. 347–354 (2005)

Author Index

Printed in the United States
By Bookmasters